# ZOMBIE
# HUMAN BIOLOGY

Stephanie J. Fischer ■ Amie Wojtyna ■ Christopher F. Green

WHAT
YOU
NEED
TO
KNOW
TO
SURVIVE
THE
ZOMBIE
APOCALYPSE

**Kendall Hunt**
publishing company

INCLUDES ACCESS TO THE **KHQ** STUDY APP

About the cover artist:

Having always enjoyed the smell and texture of paint, Billy Tackett began "splattering" with his bare fingers in order to enjoy and appreciate the organic nature of the work and eliminate brushes. The technique started out very loose with landscapes and florals but was eventually mastered and expanded into portraiture and installation pieces. Billy's work with portraiture lends itself to a complex blending of impressionism and pop art and his art in general represents a strong defiance of the traditional two-dimensional painting.

Find more about Billy Tackett's splatter work at www.billytackett.net

Cover illustration created by Billy Tackett for Christopher Green.
Hands image ©FOTOKITA/Shutterstock.com

www.kendallhunt.com
*Send all inquiries to*:
4050 Westmark Drive
Dubuque, IA  52004-1840

# CONTENTS

## CHAPTER 9: TREATMENT (OR "IF YOUR URGE TO TEAR FLESH FROM BONE AND SINEW CONTINUES FOR MORE THAN FOUR HOURS CONSULT YOUR PHYSICIAN")   174

## CHAPTER 10: PANDEMICS IN MODERN TIMES (OR…HOW I LEARNED TO STOP WORRYING AND LOVE THE VACCINE)   196

# FOREWORD

## FOREWORD BY SHAWN G. GIBBS, PhD MBA, CIH

Before you begin reading this book there is one item that we should acknowledge to make sure we all understand the point of this text—Zombies aren't real. To our knowledge, there are no secret government laboratories working on creating or defeating the zombie menace, but if such laboratories are ever created then sign us up to be the earth's last, best hope. We will hop the first available black helicopter to the secret government laboratory's undisclosed location—which will probably be in Nevada or New Mexico since it seems like that's where they would put the secret zombie laboratory—to do our best for humanity.

Since people started talking and then writing about zombies these stories have been a reflection of society, and have drawn upon scientific knowledge to make the stories as realistic as possible. So realistic are these stories that they make excellent educational tools as they both grab your attention and can impart knowledge. The Centers for Disease Control and Prevention (CDC) has even used zombies to educate the American public on how to be prepared for any emergency, using the philosophy that if you are prepared for the zombies then certainly you are prepared for an earthquake, mudslide, and anything else that Mother Nature may throw at you. It is in this spirit that we offer you this book; what better way to learn the concepts of Biology than by using our friend the zombie?

## WHAT MAKES US ZOMBIE EXPERTS?

It's not because we've seen more than our share of zombie movies, though we have. We've seen the good zombie movies (*28 days later* 2002), the great zombie movies (*Shaun of the Dead* 2004), the cult classics (*Re-animator* 1985), the movies that missed the mark (*Resident Evil* 2002), and the movies so awful they should never have been made (*Zombie Strippers* 2008). We've seen the first zombie movie, (*White Zombie* 1932), and the most influential zombie movie about 100 times (*Night of the Living Dead* 1968). We've even sat through foreign zombies like the amazing Norwegian *Dead Snow* (2009) and the amazingly bad Italian *Zombie* (1979). But at least that last one gave us the ultimate battle of zombie vs. shark (zombie wins if you're wondering). We are not zombie experts because we have seen hundreds of movies and read hundreds of books on zombies; although we clearly have.

## SCIENTISTS, EARTH'S LAST, BEST HOPE

No it's because, well we have a lot of these. That's right, we have a number of academic degrees in biology and related fields, we like zombie books and movies; and we aren't afraid who knows it. However, in full disclosure we should admit that *technically* not a single one of our academic degrees is in the field of Zombieology, but we are hoping one day this book changes that. In other words, welcome to the humble beginnings of what is sure to lead to degrees in the art of Zombie Survival.

Source: Christopher Green

We have spent years teaching biological concepts to students both in the classroom and in the laboratory, and conducting research on a wide variety of topics within biology. In our years as educators, to our surprise, we have realized that not all students enjoy biology as much as we do. In fact, there are a large number of students with little to no interest in learning about biology who are only taking biology courses to fulfill degree requirements that all graduates have a working knowledge of biological systems; these students wish to complete their biology course so that they can graduate and go on to be successful in their chosen paths. Don't worry, I'm sure that does not describe you, does it? We have prepared this

Source: Stephanie Fischer

text with those students in mind… well, at least those students who also like zombies. We wanted to impart these biological concepts to students in a way that they would hopefully enjoy, or at least enjoy more than reading your standard introductory biology textbook.

In utilizing this text book, you will gain a greater understanding of the concepts of biology as they apply to zombies, and you can then use those concepts in your daily, zombie-free life. Think you don't use biology? Think again. Every time you take a deep sniff of air to determine if there is a decaying zombie nearby—or, slightly more likely, to determine if the old take-out food in the back of your fridge is safe to eat it—then you are using biology. There are thousands of other such examples of using biology in your daily life and how you could apply it in a zombie apocalypse.

Source:

## YOU THINK YOU'RE PREPARED? THAT'S CUTE.

Is your emergency preparation kit a samurai sword and a canteen? You are probably going to accidently cut yourself and get an infection from the wound or from drinking fouled water from your canteen. Have fun being some undead fiend's lunch, unless you want to learn the biological basis for infection and how microorganisms survive in the environment. Is your emergency plan to sit on your roof with a rifle and a

cooler full of beer? Sounds fun, but again, you're on the menu. Beer will quickly dehydrate you in the open sun, and you will roll off the roof into the waiting arms of the undead; unless you want to learn the biological basis behind hydration and nutrition necessary for cellular metabolism. Survival isn't something to be taken lightly, and the odds are not in your favor. But at least by picking up this book there is a slight (*slight*) chance you might outlast the other meatsacks out there.

Of course, we are convinced we'll be picked up and taken to safety in the black helicopter the second the zombie apocalypse starts… Like I said, we're the earth's last, best hope. But good luck out there! You'll need it.

Source:

## FOREWORD BY CHRISTOPHER KIRK, ZOMBIE FANATIC

I was infected by the zombie phenomenon started back in the summer of 1989. I was up late and surfing through TV channels looking for a something to watch and I came upon a scene of a man being torn into. His intestines and organs were being strewn about and devoured by a horde of the undead.

This scene was from the 1985 Romero classic *Day of the Dead*. That particular scene always remained with me in the back of my mind because of the realism and the sheer brutality of it. Though it may have warped my young and impressionable mind somewhat, my love affair with zombies and the zombie apocalypse had begun and is still going strong 27 years later.

What is it about the shambling corpses that capture our imagination so much? Today zombies seem to be everywhere in our pop culture. From comic books and novels to TV shows, anything and everything you can imagine has been infected by the current wave of the zombie virus.

So let's talk some basic zombie history first just to fill in the blanks for you. The first time the word zombie was used in the English language was in the year 1819 by the poet Robert Southey, in the form of the word Zombi. He was referring to practices that Voodoo Bokors were said to have used potions to raise the dead and have complete control over them.

© Christopher Kirk

The first Hollywood movie to capitalize on this concept was the 1932 *White Zombie*, which in its own right is a great movie but it was still not what a modern zombie movie is. Next came the 1959 B movie classic *Plan 9 from Outer Space*, which edged us closer to the current zombie movie archetype but it was still not quite there yet as it was aliens resurrecting the dead to take over the earth.

One thing that this did was reflect the United States populace's fear of nuclear war and the spread of communism. Which I find very interesting and we will get back to that. Now let's jump ahead again to 1968, the Vietnam War is raging in south East Asia, The summer of love was in full effect and racial tensions were at an all-time high among the U.S. population and the threat of communism was very much still on everyone's mind. This was the year that zombie movies matured to its modern incarnation. The father of the modern zombie archetype George Romero released his masterpiece (*The Night of the Living Dead*)

No one had ever seen a horror movie quite like this before. It was something fresh and it reflected humanity's worst fears come to life; the taboo subject of cannibalism and that anyone could be the enemy. George Romero's intention was to reflect our countries racial disparities at the time and try to get people to think about it and have a conversation about the issue. What really got people talking were the Ghouls and the graphic cannibalism it portrayed on screen. At one point the movie was going to be called *Night of the Ghouls* but fans started referring to the creatures as zombies and the name stuck.

Really, what could possibly be more terrifying than the dead rising? That is something that mankind has been truly terrified of since before the beginning of recorded history. Then you throw in the cannibalism and you have something even more disturbing and horrifying, the average human brain cannot comprehend and sends itself directly into fight or flight mode. It's no wonder so many people love zombie movies, as most people love the adrenaline rush of being scared half to death. Zombie were to be engrained into pop culture from the foreseeable future.

Things in the Zombie movie genre changed little in the decades to follow. Zombies were still the typical shambling undead cannibals driven by the instinct to feed only. They were either reanimated by a virus or radiation and occasionally the supernatural. Then on September 11th 2001 a terrible tragedy befell America with the attack of the World Trade Center and Pentagon. This event shocked the whole country; we felt real fear for the first time together as a country since perhaps the attack on Pearl Harbor. We remembered that safety is a fallacy. With that came the war on terror and the next evolution of Zombies in pop culture. In 2002 the first of the new wave zombie movies hit with *28 days Later* and the 2004 remake of *Dawn of the Dead*.

What set these two movies apart from the others so much was how the zombies were portrayed on the screen. Long gone are the slow shambling corpses of yesterday, to be replaced by something more horrifying than ever before seen in a zombie movie. The "fast zombies", as they are called, are more akin to the raptors in Jurassic park. They move with lightning fast speed and seem far more intelligent than before. These are the Apex predators of zombies the most terrifying opponent man has fought for his survival against on the silver screen.

This was very much a reflection of our culture at the time, just like in 1968 with *Night of the Living Dead*. This seems to be a trend with zombies and our pop culture through the years. When we are scared we invent something even more terrifying than before to entertain us and take us away from the real problems of the times. As time went on, Zombies only became more and more popular. You cannot turn on the TV or go into a store without seeing Zombies everywhere.

The one thing I was always fascinated about in zombie movies was what happened to the people who survived to the end? What did they go on to do and did they survive and rebuild? Or did they eventually become a buffet for a horde of the undead? I always wondered and spent many hours thinking of whether I could survive this scenario or that one in the event of a zombie apocalypse. I myself have consumed and digested every possible zombie movie and book on this subject only to be left with the same standing question—could I survive?

Could you survive? Do you have the Knowledge? Do you have the Skills? My advice to you is use the information in this book to help yourself survive the impending zombie apocalypse.

# ACKNOWLEDGMENTS

The authors would like to thank the following individuals for
their continued support of this project:

Emily Adele, contributing artist
Michelle L. Bahr, Kendall Hunt Publishing Company
Blake Bextine, PhD
Damon Daugherty
Ego, contributing artist
Belinda Fischer
Shawn G. Gibbs, PhD, contributing author
Xander William Green
Christopher Russell Kirk
Aaron Lambert, contributing artist
Karen Mathis, PhD, contributing editor
Michael "Krom" McNeese, contributing editor
Travis Parsons, SSgt USAF, contributing author
Aimee Schmidt, Kendall Hunt Publishing Company
Billy Tackett, contributing artist

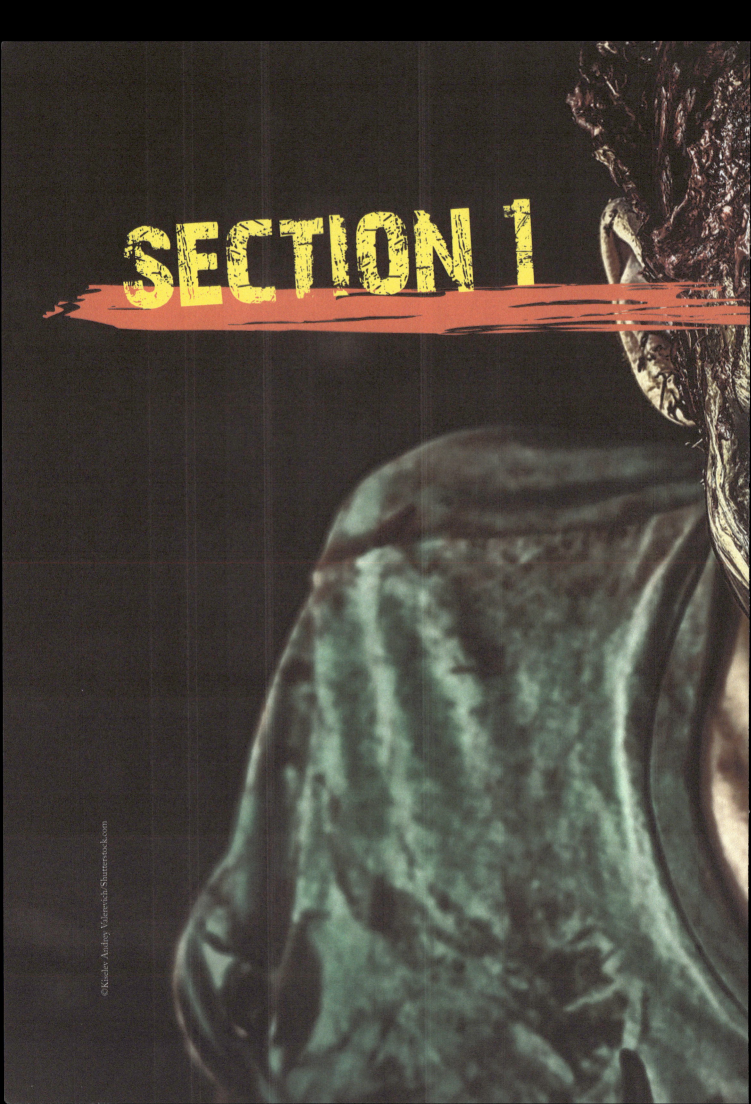

# SECTION 1

**Welcome to the Zombie Apocalypse. Hope You Survive the Experience.**

# CHAPTER 1

## THE BASICS

**(OR... YOU GOTTA LEARN TO WALK BEFORE YOU RUN OR SHAMBLE)**

You've been warned for years to prepare yourself, but you scoffed at the notion of a zombie infestation. Well, you should have listened, because it's no longer science fiction. It's finally happened! An outbreak! Scattered, often conflicting reports are coming in over the airwaves. So many questions are going through your head.

1. Where can I go for help?
2. Where will I turn for information?
3. What is causing the outbreak? Is it genetic? a mutation? a chemical? some sort of microorganism?
4. How is it spreading? How can I keep my family from becoming infected?
5. How can I save my loved ones?

All these questions and more will be answered by the end of the book. If you're one of the lucky ones who knows this information before an outbreak has occurred, then this might just save you and yours from being zombie food. *Might*.

## WHERE CAN YOU FIND RELIABLE INFORMATION?

Where can you go for help? Where can you find reliable information? This is a question for people every single day… from figuring out what is causing your sore throat to understanding a biopsy result and planning your chemotherapy.

With the advent of the World Wide Web and social media, people are bombarded daily with a glut of information. Unfortunately, this has led to more and more people accessing unreliable medical information and sometimes risking their lives—as well as the lives of loved ones—based upon fictional medical stories. In the case of a true zombie outbreak, who can you trust? It turns out there are some reliable sources for medical information, and if you know where to access it, you are already well ahead in the game of survival. Reliability is composed of several key components:

- Who is doing the talking?
- What are their qualifications?
- What is their agenda (that is, what are they trying to achieve by convincing you)?

## SCIENTIFIC METHOD AND PUBLISHING SCIENTIFIC WORK

The scientific method plays a large role in eliminating bias in scientific and medical studies. There is a format for how a scientific or medical work is published, and it reflects the scientific method.

The scientific method involves first educating yourself on what has already been done on your topic of interest, and then coming up with a hypothesis. A hypothesis is a testable statement that you can then specifically design an experiment to test.

Designing the experiment carefully is very important… a key part of the method is that, if at all possible, only one condition (variable) should change, and everything else should be kept as constant as possible. In addition, you should include a control group, where the variable does not change, to compare your results.

Once you get your results, you should repeat your experiment and analyze the results to find out if the results consistently altered in a similar way, and run statistical tests to find out whether the results could have happened by chance (even if your hypothesis was not true).

Scientific work can be published as a poster (often presented at a conference of scientists or physicians) or as a paper (published in a peer-reviewed scientific journal). Both follow the same format:

- Title: The title of the scientific paper or poster should be specific.
- Abstract: The abstract is generally a paragraph long and summarizes the entire experiment. The objective, experimental design, results, and conclusions are all included.
- Introduction (*or* Background): The introduction will introduce the reader to the topic, often explain why the study is important, and describe the objectives of the experiment or study. The introduction will also include the hypothesis.
- Experimental Design (*or* Materials and Methods): The experimental design section describes how the experiment was performed and provides information about how the data were collected.
- Results: The results section summarizes the data collected and/or observations made. No conclusions are made in this section. Interpretations of the data occur in the next section.
- Discussion: In this section the results are interpreted and conclusions drawn. The author(s) try to make sense of the data. Often, flaws in the experiments are discussed with ways to improve the experiment. Finally, the hypothesis is accepted, modified, or rejected in this section.
- Literature Cited (*or* References): Throughout the paper or poster the author(s) may cite other papers and experiments. At the end of the paper, the Literature Cited section provides the full reference for the reader.

The most reliable science is done with very high experimental numbers and very controlled circumstances. That can be really difficult to achieve in humans (we all eat so differently, are so different genetically and live in a different environments). Many times we do our basic science in animal models, where we can use larger numbers and get more consistency in the treatment groups, before trying things on humans. Just to clarify, most scientists LOVE animals, and do this work with the utmost compassion for the animals. Scientists follow a principle called "the 3 Rs": *reduce* the number of animals used to the absolute minimum, *refine* experiments to be the kindest possible to the animals, and *replace* the use of animals with tissue culture and other models as much as possible. However, while animal-based research is something we would all like to minimize as much as possible, it has undeniably contributed to significant improvement in the medical field. The list of treatments and therapies that are a direct result of animal research is staggering. HIV/AIDS, diabetes, cancer, and heart disease are just a few examples of the multitudinous diseases in which breakthroughs have been achieved due to animal-based research.

A key part of a well-designed scientific experiment is that it is repeatable. Scientists should write their methods clearly so that anyone can repeat their experiments, and they should repeat their own experiment multiple times so they know their results are consistent and reliable. Scientists analyze their data (see how different their results are for each individual test animal or condition) using statistics to determine if differences in results are statistically valid or simply the product of chance.

## PEER-REVIEWED SCIENCE

Most medical science in the United States is performed by scientific laboratories or medical hospitals that are funded (at least in part) by taxpayer dollars. This means that you can have access to the information at reliable websites and agencies. Usually those websites are summarizing a published scientific poster from a science conference or a published scientific paper from a peer-reviewed journal.

 FIGURE 1.1.

Public health care agencies.

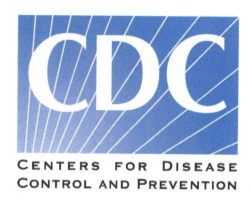

CENTERS FOR DISEASE
CONTROL AND PREVENTION

© yui/Shutterstock.com

When a scientist or a physician designs an experiment, they have to make sure the scientific method is used. Remember they are trying to eliminate bias, and other problems that might interfere with gathering reliable data for a defined question. This means that there has to be:

- Good controls, so that only one variable is changed at a time. For example, if you are testing a new anti-zombification medicine, you need to test it on a group of zombie animals that are all of similar age and medical background (like, they've all been zombies for the same amount of time, and were exposed from the same source). The scientists design the experiment so there are good controls (one group of zombie animals that is not getting the new medicine, to compare to).
- The same delivery method (injection into zombie muscles, delivery in food, pills, drinks?).
- A scientific way to measure improvement in zombie symptoms (measuring heart rate in the zombies, as opposed to asking how the zombie feels).
- Repetitions of the experiment several times in different groups of zombies and statistical analysis to make sure the result is not simply due to random chance.
- Finally, when you report your results, you need to say exactly what happened, and not make leaps of logic. For instance, you can say that your new drug treatment over 14 days helped to improve heart rates in zombie rabbits who had been zombified by bite at time zero. However, you can't say your new drug is a cure for zombies.
- Clinical human studies must be ethical, must have educated the participants fully, cannot withhold effective treatment, and are trickier to report because humans have so many variables that are hard to keep constant.

## High doses of caffeine to restore heart rate in new human zombies

S Fischer, C Green, S Gibbs, and T Parsons

One of the early symptoms of zombieism is the slowing and ultimate loss of the heart beat. Previous work has demonstrated the ability of caffeine to increase heart rate, and caffeine is easily administered to new zombies.

Heart Rate measurements were taken via finger pulseometer every 2 hours (prior to consumption of caffeine) for 2 weeks of caffeine exposure and for the following 6 weeks.

Results were analyzed by ANOVA, experiment was repeated 2 more times with 60 more zombies over a four month period.

We hypothesized that if we administered high doses (equivalent to 5 cups of coffee) of caffeine to recent zombies, then we would observe increased heart rate and potentially delayed zombie symptoms.

Data demonstrate caffeine maintains incresed heart rate in patients bitten by zombies for several weeks.

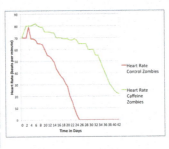

P values were significant at all time points up until 3 and a half weeks, p < 0.05. At that point SE increased to non-significance in all test groups.

Materials and methods: We recruited 60 newly exposed zombies (bitten within the previous 14 days) in the Tyler TX area. Half were randomly assigned to the control group (receiving placebo), and half received 500 mg caffeine in liquid form (100 mg every 2 hours during waking hours) for 2 weeks.

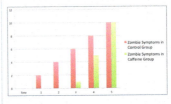

Our hypothesis was supported, in that high doses of caffeine administered to new zombies for 2 weeks did increase heart rate for an additional 2 weeks. However, by the fourth week all zombies had falling heart rates and were exhibiting other zombie symptoms. Our caffeine treatment did preserve heart rate but did not prevent zombification in patients.

Source: Stephanie Fischer

Such scientific and clinical studies are published by the authors in peer-reviewed journals as a scientific paper. That means that other scientists and physicians have already read the reports and have said that the results are plausible, the methods were good, and the conclusions are reasonable.

Sometimes bad science gets through the peer-review process. For instance: a 1998 paper in the British journal *The Lancet* purported a link in the administration of the measles, mumps, and rubella (MMR) vaccine to autism (Wakefield, 1998). After a 2004 investigation, *The Lancet* editors issued a response disputing the claims (Horton, 2004) before retracting the original article. Eventually, the UK review board cited deliberate falsification in the research published in *The Lancet* (i.e., he just made it up!!!) and barred the now former surgeon and researcher from practicing medicine. But oh, we'll talk about this joker in much more detail later.

# NUMBERS DON'T LIE, DO THEY?

Oh you poor, naïve future zombie snack. Numbers lie all the time. If others want to lie to you with numbers, the easiest way for them to do it is to put the numbers in graph form. The next time someone on television shows you a graph, keep in mind the components of reliability mentioned earlier. Say a TV personality is arguing that crime rates are up in New York City. To prove this the person shows you this frightening graph:

**FIGURE 1.3.**  Not quite graph, New York City crime rates.

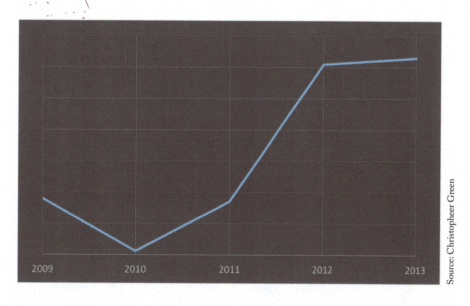

Now what is this graph actually telling us? Well… nothing. Often graphs shown on TV lack even basic components necessary to convey useful information. So let's start there. We can first add the title, axis numbers, and axis titles. Let's also find out where they got that data.

FIGURE 1.4.

Looks legit to me.

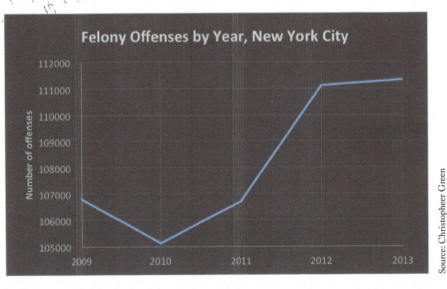

Source: Christopheer Green

Well that's a good start. Wow, what a scary graph! New York is clearly a war zone. How does anyone survive in that savage wasteland? Since we now have all the components in our graph, it's cut and dried, right? But where did they get this data? Well, if we look at the source material the graph might look a little different.

FIGURE 1.5.

Felony offenses by year, New York City.

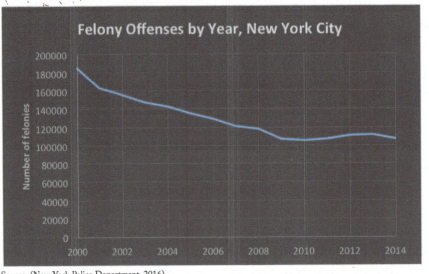

Source: Christopheer Green

Source: (New York Police Department, 2016)

The source actually has data from 2000 up to 2014. Also look at the scale on the X axis. The previous scale only showed seven percent of the total scale. It is unfortunately an all too common practice to modify the scale of the axes to skew the data to serve an agenda. The last graph with the complete data set shows the full numerical scale. Notice the full scale and data set show quite a difference between what the TV personality wanted you to see and the actual numbers. Finally keep in mind that a more accurate assessment of felony statistics would take into account total population to calculate the felony rate per capita (for example, the NYC population in 2014 was just shy of 8.5 million).

## SPIN

One big problem right now is that a lot of the revenue ($$) on big media company websites (like cable channels, websites, big newspapers, and even Facebook) is dependent on how many people click on a link. Basically, an advertiser like *Zombie-Cola*\* says "we will pay you this month based on how many people read your website, and we are going to measure that by measuring how many people click on your articles on Saturday December 13."

So media companies REALLY need people to click on links to read articles. That means they have hired people to design "click bait" links.... links that give you just a hint of what the story might be about, to make sure you click and get counted towards their advertiser dollars.

**FIGURE 1.6.** Billionaire playboy with a penchant for theatrics.

Too often, what these media companies are doing are taking peer-reviewed scientific articles and twisting their words, or making leaps of logic that aren't actually addressed in the scientific paper.

An example that you might be familiar with is a movie... a click bait article might lead with the following:

Billionaire playboy commits brazen assaults, leads to numerous deaths.

©Olga Popova/Shutterstock.com

There is a whole lot more to *The Dark Night* than that, and it's not really an accurate reflection of the plot. But maybe not a huge leap of logic there... at least the headline is saying that *one particular* billionaire playboy committed assaults, not that *all* billionaire playboys are committing assault. Sometimes media companies aren't even that careful...

Currently *not* in contention for father of the year.

©Stefano Buttafoco/Shutterstock.com

That is a prime example of taking a very limited study and claiming that it applies to a large group of patients. Certainly Darth Vader (and Grand Moff Tarkin) destroyed the entire planet of Alderaan along with Princess Leia's entire family, but there's no evidence to suggest that this is a common occurrence. That's an example of overgeneralization, which happens way too often when media reports on scientific studies.

So, too often, what you read on a news site isn't a very accurate report of what the scientists actually said.

On another note, if you don't know what movie the above click bait is referring to, this textbook is clearly not for you.

Spin is also often used on legitimate scientific research by news outlets. Maybe, after some animal trials were done, someone was able to recruit some actual human zombies to test on. Usually human tests won't be as clear-cut as basic science studies with cells or animals; again, humans tend to be pretty uncontrollable in a lot of ways (diet, behavior, environment) but humans do have the advantage that you can ask them how they are doing, feeling, etc. Let's think about how the media might spin the result, and how that might affect people who don't know how to find or read reliable scientific studies.

As an example, let's revisit figure 1.3. This poster shows that experimentally, high doses of caffeine given to newly exposed zombies was able to maintain a higher heart rate over 6 weeks, but that ultimately this did not prevent other zombie symptoms. Caffeine did not prevent the exposed people from becoming zombies.

However some media outlets would not hesitate to write this up as: "Caffeine slows zombie conversion! Click here to see new study!", or even...

...and certain people who take money from companies (let's say, in this case, the ultra-caffeinated *Zombie-Cola\**) would go on their TV shows or Facebook pages and say "Dr. Blue interviews zombies cured by caffeine drinks! See more tonight."

It wouldn't take long for some celebrity out there to come out and say that they always knew that caffeine was really good for you, and this proves that pregnant women should drink lots of caffeine during pregnancy to protect their kids from becoming zombies...

Remember, the scientists *said no such thing*.

*\*Zombie-Cola* is NOT a real company.

## RELIABLE SOURCES FOR SCIENTIFIC AND MEDICAL INFORMATION

Where are some good reliable places to go to find science that has been made "digestible" for the average person?

One such source is sciencedaily.com. In addition to summarizing the peer-reviewed articles, this site also links to the peer-reviewed articles so you can find them and read them yourself. The people there work hard to report accurately and not make leaps of logic. They can also give good background information. Check it out!

Now what about a reliable source for medical information? Let's say you've just taken your daughter to the pediatrician, who tells you she has been bitten and is becoming a zombie. Where do you go for accurate information about what new treatments might be out there?

There are a couple of reliable places that scientists and doctors go, and you can see them, too.

There is a group of "Core Clinical" journals who have reliable peer-review processes and are known to make sound editorial decisions (as opposed to journals who charge people to publish in them but accept any article without reading it, or journals who publish what their government tells them to publish). This collection is called the "Abridged Index Medicus", and you can find a list of these journals at **nlm.nih.gov/bsd/aim.html**. Other reliable sources include **MedlinePlus.gov**, Johns Hopkins Medicine Health Library, and **MayoClinic.com**.

Let's say you're a patient who has been diagnosed with zombieism, and you want to find a clinical trial where they are testing drugs on willing and informed patients… where do you go to find out if anyone is offering that kind of clinical trial? One of the best places to go is **ClinicalTrials.gov**. There you can search by disease or by state to find a clinical trial near you that you might be able to participate in. There are usually restrictions, like age or stage of disease, but it is a good resource to ask your doctor about.

**MOVIE SPOTLIGHT:** *Shaun of the Dead* (Wright, 2004)
Causative agent – Though it's hinted at several times, the causative agent is never identified.
A down-on-his-luck slacker needs to save his ex-girlfriend, his friends, and parents from a zombie outbreak that has decimated London.
"Take car. Go to Mum's. Kill Phil – 'Sorry.' - grab Liz, go to the Winchester, have a nice cold pint, and wait for all of this to blow over. How's that for a slice of fried gold?"
"Yeah, boyyyeee!"

# CHAPTER 1    QUESTIONS/WORDSTEMS

1. The scientific method is used to make certain experiments are (choose all the correct answers)
   a. Reliable (produces the same result when repeated).
   b. Repeatable by other scientific groups.
   c. Important to everyone.
   d. Funded by government.
   e. Everyone who reads it should be able to understand it immediately.
   f. Unbiased by those performing the experiment.

2. Which is MOST important when evaluating scientific information that you have heard or read about? (choose one)
   a. The credentials of the source.
   b. The format of the information (television vs. journal vs. news).
   c. How much money the study cost.
   d. How many other people have read it.
   e. How famous the person presenting the information is.

3. True or False: Scientists are required to tell everyone if they are being paid by a company that makes a product tested in an experiment (or a competitor's product). Justify your answer.

4. True or False: Celebrities and reporters are required to tell everyone if they are being paid by a company that makes a product they are recommending or talking about. Justify your answer.

5. Standard Error, Standard Deviation, and p values are all important parts of scientific analysis because they
   a. Indicate how consistent, reliable, and significant the results are.
   b. Indicate how easily the results may be translated into humans.
   c. Indicate how expensive the experiment was.
   d. Make the experiment very repeatable.

| WORDSTEMS | |
|---|---|
| bio- | life |
| chemo- | chemical |
| hypo- | under; beneath |
| -logy | study of |
| qualita- | degree of difference by comparison |
| quantita- | having a mesaurable magnitude or difference |
| -thesis | arranging |
| -therapy | treatment |
| var- | different |

# CHAPTER 1 — WORKSHEET

## STATISTICS AND GRAPHING (GROUP PROJECT)

Adapted with permission from (Green, Clark, Mathis, Barkhurst, & Mansfield, 2016)

## OBJECTIVES

1. Distinguish between qualitative and quantitative data
2. Calculate statistics (mean, median, and standard deviation) of given data
3. Given a data set, create a column and scatter graphs using Excel

## INTRODUCTION

Data is defined as, "factual information (as measurements or statistics) used as a basis for reasoning, discussion, or calculation" (Merriam-Webster, 2012).[i] Data can be either quantitative (measurable) or qualitative (dealing with descriptions). Examples of quantitative data include measurements such as weight, height, blood pressure, and heart rate. Qualitative data would be observations such as taste, color, smell, and appearance. Data can be summarized by performing statistics (mathematical calculations) on the numbers. Some common statistical calculations used in data analysis are the mean (or average), median, and standard deviation. The mean is calculated by summing the individual values and dividing by the number of values in the data set. The median is the midpoint between the highest and lowest values in the set. The standard deviation describes how spread apart from the average the values are. The equation for standard deviation is as follows:

Where

$$\sigma = \sqrt{\frac{\sum_{i=1}^{N}(x_i - \bar{x})^2}{n-1}}$$

$\sigma$ = standard deviation
$\Sigma$ = sum of
$\overline{n}$ = numbers of values in the data set
$\bar{x}$ = mean

Data can be summarized and displayed using tables and figures. One figure commonly used is the graph. A graph is a visual representation of the data that allows a reader to easily see the results and trends in an experiment. Graphs have many forms including pie charts, line graphs, and column graphs. Graphs can be constructed by hand (using graph paper); however, with the accessibility and ease of computer programs such as MS Excel, graph construction proceeds more quickly and with a neater result. When graphing, it is important to differentiate between variables, data you are trying to measure. Independent variables stand alone and aren't affected by other factors, while dependent variables are affected as the name implies. When building the graph, independent variables will be on the X axis while dependent variables will be on the Y axis.

Each graph that you construct must have the following information:

1. Chart title  (descriptive of the data collected)
2. X axis label (including the units used)
3. Y axis label (including the units used)
4. Legend (if graphing two data sets on one graph)
5. Data

## PROCEDURE

### Part A: Data and Statistics

- Define each of these data as *qualitative* or *quantitative*.
    a. These zombies have a more pungent odor than most.          _____
    b. This zombie has a top speed of 4 MPH.                      _____
    c. These brains have a strong taste.                          _____
    d. I have fought off 8 zombies.                               _____

- Use MS Excel to calculate the mean, median, and standard deviation for the following test scores: 80, 70, 54, 67, 52, 75, 93, 86, 73, 88, 91, 79, 61, 77, 82.

## Part B.  Making Graphs

1. Create a column graph in MS Excel of the following data. Determine which of the variables is independent (for the X axis) and which is dependent (Y axis). Paste the graph in the space provided.

**TABLE 1.1**                Number of infected in select major U.S. cities.

| City | Estimated number of infected |
|---|---|
| Milwaukee, WI | 5,000 |
| Cleveland, OH | 90,000 |
| Chicago, IL | 150,000 |
| Lexington, KY | 15,000 |

2. Create a column graph in MS Excel of the following data. Determine which of the variables is independent (for the X axis) and which is dependent (Y axis). Paste the graph in the space provided.

**TABLE 1.2**    Zombie infection rates (confirmed cases per 100,000 persons).

| Date | Ohio | Kentucky |
|------|------|----------|
| 09/12/2016 | 0.1 | 0.0 |
| 09/13/2016 | 0.2 | 0.1 |
| 09/14/2016 | 0.3 | 0.1 |
| 09/15/2016 | 0.3 | 0.1 |
| 09/16/2016 | 0.3 | 0.1 |
| 09/17/2016 | 0.4 | 0.1 |
| 09/18/2016 | 0.4 | 0.2 |
| 09/19/2016 | 0.4 | 0.2 |
| 09/20/2016 | 0.5 | 0.4 |
| 09/21/2016 | 0.6 | 0.4 |

3. Create a scatter graph (with line) in MS Excel of the following data. Determine which of the variables is independent (for the X axis) and which is dependent (Y axis).
Brian was bitten during the last attack on the compound. He was rushed to the field hospital where his vitals were measured immediately and every 30 minutes thereafter.

**TABLE 1.3**                                    Patient heart rate (BPM) after infection.

| Time (minutes) | Heart rate (beats per minute) |
|---|---|
| 0 | 73 |
| 30 | 78 |
| 60 | 89 |
| 90 | 92 |
| 120 | 95 |
| 150 | 110 |
| 180 | 130 |
| 210 | 155 |
| 240 | 180 |
| 270 | 0 |

# CHAPTER 2

## DEAD OR ALIVE

**(OR… "CHECK HIS PULSE IF YOU WANT… I'M RUNNING!")**

Attach to toe

Name of Deceased

Age | Sex | Race

Place of Death

Cause of Death

Physician

Funeral Director

Comments

So this… person is approaching you. Not quite walking… but stumbling and falling toward you. There's no other way to put it… he looks dead; but if he's dead how can he be walking around? You don't want to use the word zombie… but how else would you describe this thing coming toward you with dead eyes, snarling, salivating, teeth exposed, clearly intent on violence?

So before you grab your trusty _____ (shotgun? baseball bat? Samurai sword? This is not that kind of book!), take just a moment to ask yourself one fundamental question: Is this person really dead? How can a zombie be technically dead, but still walking around? What exactly do we mean when we say something is "dead" vs. "alive"? If he isn't alive, are you technically killing him? Keep in mind when we say take a moment, make it a quick moment. There's still a zombie bearing down on you.

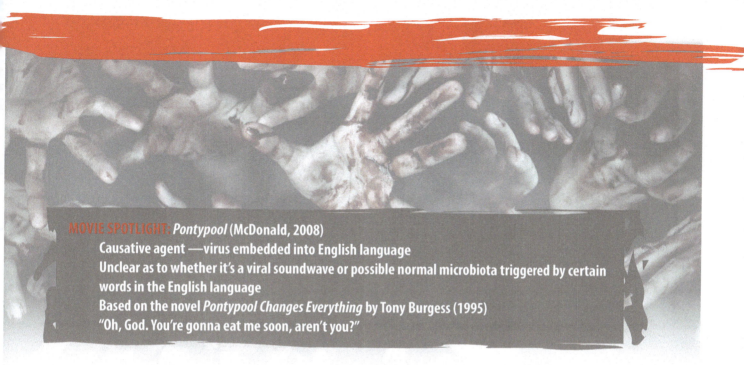

# WHAT MAKES A ZOMBIE? THE CHARACTERISTICS OF ALL LIVING ORGANISMS

We know biology is the study of life. The most fundamental question for biology is, of course, what is life? Scientists define seven distinct characteristics that all living things share (Table 2.1). We will talk about each of these characteristics throughout the book. Finally at the end of the book we will answer the fundamental question: *are zombies truly dead?*

TABLE 2.1          Seven characteristics of all living things.

- Living things are highly ordered.
- Living things obtain energy from their environment for metabolism.
- Living things sense and respond to their environment.
- Living things maintain homeostasis.
- Living things grow and develop.
- Living things reproduce via DNA.
- Living things evolve.

## LIVING THINGS ARE HIGHLY ORDERED

The first cells were viewed by scientist Robert Hooke in the mid-1600s. He was looking through early simple microscopes at cork. He was the first to coin the term "cell" after observing the honeycomb appearance of the plant tissue. By 1838 botanist Matthias Schleiden and physiologist Theodore Schwann had come up with a theory to describe how cells relate to living things. It is called the "cell theory" and is summarized as follows:

1. All living organisms are made of one or more cells. Bacteria, protozoans, certain fungi, and algae consist of just a single cell. Larger organisms (such as humans) are known as multicellular organisms.
2. The cell is the most basic unit of life. All cells have cytoplasm housed behind a plasma (cell) membrane to hold in everything.
3. All cells arise from preexisting, living cells.

Taken together, these three tenets describe the theory of how cells are the key component of living things.

We live because our cells live, so to speak. Figure 2.1 shows a typical plant and animal cell with labeled organelles ("little organs") that have specialized functions in the cell. Furthermore, all living things are ordered into one or more complex cells, the smallest structures of life.

All cells are highly organized with a plasma membrane, made up of lipids and proteins, which surrounds the contents of the cell. Simpler bacterial cells are called prokaryotes and have minimal components inside the plasma membrane. More complex cells, including animals, plants, fungi, algae, and protozoans, are called eukaryotes. They are 10x to 100x larger than prokaryotic cells and have the highly structured organelles pictured in Figure 2.1. For instance, the nucleus houses cut your genetic information while the mitochondrion generates energy from large organic molecules via cellular respiration. Table 2.2 lists the function of select organelles.

**TABLE 2.2**

Function of select Eukaryotic cell organelles.

| | |
|---|---|
| **Cytosol** | Viscous inner liquid of the cell. |
| **Nucleus** | Houses the genetic information of the cell (DNA) in chromosomes. |
| **Mitochondrion** | Synthesis of ATP via cellular respiration. |
| **Ribosomes** | Responsible for translation, production of protein from mRNA. Can be found on the rough ER or floating in the cytosol. |
| **Rough Endoplasmic Reticulum (RER)** | Covered in ribosomes; site of protein synthesis. |
| **Smooth Endoplasmic Reticulum (SER)** | Multiple functions, including synthesis of lipids and the breakdown of toxins in the liver. |
| **Golgi Apparatus** | Modifies, sorts, and packages cell products, ships them to their destination inside or outside of the cell. |
| **Lysosomes** | Intracellular digestion and recycling of unwanted chemicals. |
| **Cytoskeleton** | Found only in animal cells; provides cell shape, organizes the internal cell; transports substances throughout the cell. |
| **Chloroplasts** | Found only in plant cells; synthesize glucose via photosynthesis. |
| **Vacuole** | Found only in plant cells; maintains turgidity in the plant cell to make a more rigid structure. |
| **Cell Wall** | Found only in plant cells; thick cellulose layer outside the cell membrane that protects the plant cell and adds to its rigidity. |

FIGURE 2.1.

Typical animal and plant cells.

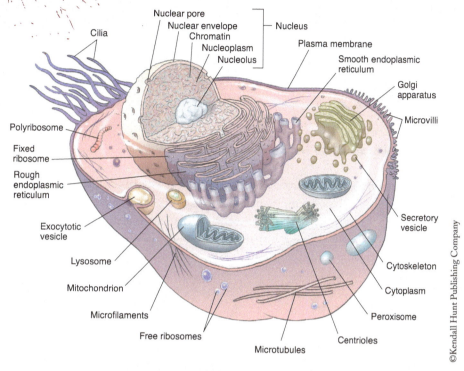

Cilia

Nuclear pore
Nuclear envelope
Chromatin
Nucleoplasm
Nucleolus

Nucleus

Plasma membrane

Smooth endoplasmic reticulum

Golgi apparatus

Microvilli

Polyribosome

Fixed ribosome

Rough endoplasmic reticulum

Exocytotic vesicle

Lysosome

Mitochondrion

Microfilaments

Free ribosomes

Microtubules

Secretory vesicle

Cytoskeleton

Cytoplasm

Peroxisome

Centrioles

©Kendall Hunt Publishing Company

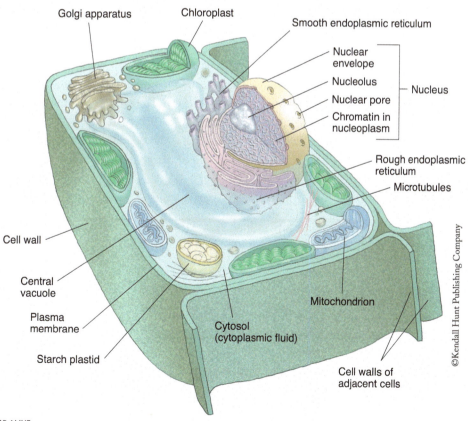

Golgi apparatus

Chloroplast

Smooth endoplasmic reticulum

Nuclear envelope

Nucleolus

Nuclear pore

Chromatin in nucleoplasm

Nucleus

Rough endoplasmic reticulum

Microtubules

Cell wall

Central vacuole

Plasma membrane

Starch plastid

Cytosol (cytoplasmic fluid)

Mitochondrion

Cell walls of adjacent cells

©Kendall Hunt Publishing Company

## LIVING THINGS OBTAIN ENERGY FROM THEIR ENVIRONMENT FOR METABOLISM

The processes cells use to convert energy for survival is referred to as a cell's metabolism. Animal cells, for instance, break down large organic molecules that they consume from other organisms such as carbohydrates, proteins, and lipids to generate energy. Plants, in contrast, transform energy from sunlight into useable energy and to build thick protective cell walls. Cells use biological catalysts called enzymes to build and break bonds between atoms. Bonds take energy to form and release energy when they are broken. Cells are effectively storing energy in larger molecules. Then when the cells break down the larger molecules, small packets of energy are released.

## LIVING THINGS SENSE AND RESPOND TO THEIR ENVIRONMENT

Living things can gather information from their surrounding environment and respond. Such response is called behavior. Animals such as humans have obvious ways of detecting stimuli such as sight, smell, hearing, etc., but all organisms respond to the environment in some fashion. For instance, plants grow towards sunlight and sense gravity. Many prokaryotic bacteria, the simplest organisms on the planet, can swim towards food or sunlight depending on their needs.

## LIVING THINGS MAINTAIN HOMEOSTASIS

The plasma membrane that surrounds all living cells acts as a selectively permeable membrane to help separate the inside of the cell from the surrounding environment. This allows cells to keep internal conditions constant even if external conditions are always changing. This maintenance of constant internal conditions is known as homeostasis and is a fundamental tenet of biology.

The phospholipid bilayer shown in Figure 2.2B stops most molecules from entering or exiting the cell. Only molecules like $O_2$ and $CO_2$ (small, nonpolar molecules with no charge) can move in and out of the cell freely. Your cell can *selectively* decide what else it wants in or out of the cell. Everything else the cell wants in or out has to go through integral proteins that the cell can open or close depending on its needs. This ensures the chemical constituency inside of the cell is different from the outside. This is the fundamental process of homeostasis.

## LIVING THINGS GROW AND DEVELOP

All living things grow in size and complexity once they arise from existing organisms. Some grow to a maximum size at maturity while some never stop growing.

## LIVING THINGS REPRODUCE VIA DNA

As stated earlier, all cells arise from preexisting, living cells. Prokaryotic bacterial cells arise from a process known as binary fission, a form of asexual reproduction. Larger eukaryotic multicellular organisms such as plants and animals undergo sexual reproduction. The offspring of these organisms get their characteristics

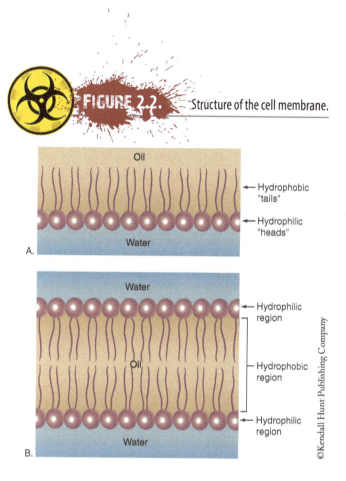

©Kendall Hunt Publishing Company

**FIGURE 2.2.** Structure of the cell membrane.

from their parents via genetic information called deoxyribonucleic acid (DNA). DNA stores all the information of an organism's physical and physiological traits as well as behavior.

## LIVING THINGS EVOLVE

Evolution refers to the change in the genetic traits from one generation to the next. Natural selection is the mechanism that changes these genetic traits, which arises from stresses in the environment. In a genetically diverse population (group of a single species in an area), certain organisms will possess advantageous traits which favors their survival and reproduction over others. This advantage will cause adaptation of the species as a whole to the environment. The chemistry part is fantastic—love it.

## A (VERY) BRIEF OVERVIEW OF CHEMISTRY

Ugh... I know, I know, you don't want to learn chemistry. You took great pains to avoid it and here we are, sneaking it into a biology book. But you can't really understand the biology of zombies without at least a basic understanding of what's happening at an atomic level. So like pulling a Band-Aid, the best way is to just get it over with.

Sigh... sorry in advance.

FIGURE 2.3.

The periodic table.

| Group IA | | | | | | | | | | | | | | | | | | VIIIA |
|---|---|---|---|---|---|---|---|---|---|---|---|---|---|---|---|---|---|---|
| Hydrogen 1 **H** 1.0079 | IIA | | | | | | | | | | | | IIIA | IVA | VA | VIA | VIIA | Helium 2 **He** 4.0026 |
| Lithium 3 **Li** 6.941 | Beryllium 4 **Be** 9.0122 | | | | | | | | | | | | Boron 5 **B** 10.811 | Carbon 6 **C** 12.0112 | Nitrogen 7 **N** 14.0067 | Oxygen 8 **O** 15.9994 | Fluorine 9 **F** 18.9984 | Neon 10 **Ne** 20.179 |
| Sodium 11 **Na** 22.989 | Magnesium 12 **Mg** 24.305 | IIIB | IVB | VB | VIB | VIIB | | VIIIB | | IB | IIB | | Aluminum 13 **Al** 26.9815 | Silicon 14 **Si** 28.086 | Phosphorus 15 **P** 30.9738 | Sulfer 16 **S** 32.064 | Chlorine 17 **Cl** 35.453 | Argon 18 **Ar** 39.948 |
| Potassium 19 **K** 39.098 | Calcium 20 **Ca** 40.08 | Scandium 21 **Sc** 44.956 | Titanium 22 **Ti** 47.90 | Vanadium 23 **V** 50.942 | Chromium 24 **Cr** 51.996 | Manganese 25 **Mn** 54.938 | Iron 26 **Fe** 55.847 | Cobalt 27 **Co** 58.933 | Nickel 28 **Ni** 58.71 | Copper 29 **Cu** 63.546 | Zinc 30 **Zn** 65.38 | Gallium 31 **Ga** 69.723 | Germanium 32 **Ge** 72.59 | Arsenic 33 **As** 74.922 | Selenium 34 **Se** 78.96 | Bromine 35 **Br** 79.904 | Krypton 36 **Kr** 83.80 |
| Rubidium 37 **Rb** 85.468 | Strontium 38 **Sr** 87.62 | Yttrium 39 **Y** 88.905 | Zirconium 40 **Zr** 91.22 | Niobium 41 **Nb** 92.906 | Molybdenum 42 **Mo** 95.94 | Technetium 43 **Tc** (99) | Ruthenium 44 **Ru** 101.07 | Rhodium 45 **Rh** 102.905 | Palladium 46 **Pd** 106.4 | Silver 47 **Ag** 107.868 | Cadmium 48 **Cd** 112.40 | Indium 49 **In** 114.82 | Tin 50 **Sn** 118.69 | Antimony 51 **Sb** 121.75 | Tellurium 52 **Te** 127.60 | Iodine 53 **I** 126.904 | Xenon 54 **Xe** 131.30 |
| Cesium 55 **Cs** 132.905 | Barium 56 **Ba** 137.34 | 57 | Hafnium 72 **Hf** 178.49 | Tantalum 73 **Ta** 180.948 | Tungsten 74 **W** 183.85 | Rhenium 75 **Re** 186.2 | Osmium 76 **Os** 190.2 | Iridium 77 **Ir** 192.2 | Platinum 78 **Pt** 195.09 | Gold 79 **Au** 196.967 | Mercury 80 **Hg** 200.59 | Thallium 81 **Tl** 204.37 | Lead 82 **Pb** 207.19 | Bismuth 83 **Bi** 208.980 | Polonium 84 **Po** (209) | Astatine 85 **At** (210) | Radon 86 **Rn** (222) |
| Francium 87 **Fr** (223) | Radium 88 **Ra** (226) | 89 | Rutherfordium 104 **Rf** (261) | Hahnium 105 **Ha** (262) | Seaborgium 106 **Sg** (263) | Neilsbohrium 107 **Ns** (264) | Hassium 108 **Hs** (265) | Meitnerium 109 **Mt** (266) | | | | | | | | | |

Lanthanides

| Lanthanum 57 **La** 138.91 | Cerium 58 **Ce** 140.12 | Praseodymium 59 **Pr** 140.907 | Neodymium 60 **Nd** 144.24 | Promethium 61 **Pm** 144.913 | Samarium 62 **Sm** 150.35 | Europium 63 **Eu** 151.96 | Gadolinium 64 **Gd** 157.25 | Terbium 65 **Tb** 158.925 | Dysprosium 66 **Dy** 162.50 | Holmium 67 **Ho** 164.930 | Erbium 68 **Er** 167.26 | Thulium 69 **Tm** 168.934 | Ytterbium 70 **Yb** 173.04 | Lutetium 71 **Lu** 174.97 |
|---|---|---|---|---|---|---|---|---|---|---|---|---|---|---|

Actinides

| Actinium 89 **Ac** | Thorium 90 **Th** | Protactinium 91 **Pa** | Uranium 92 **U** | Neptunium 93 **Np** | Plutonium 94 **Pu** | Americium 95 **Am** | Curium 96 **Cm** | Berkelium 97 **Bk** | Californium 98 **Cf** | Einsteinium 99 **Es** | Fermium 100 **Fm** | Mendelevium 101 **Md** | Nobelium 102 **No** | Lowrencium 103 **Lr** |
|---|---|---|---|---|---|---|---|---|---|---|---|---|---|---|

Name - Hydrogen
Atomic # - 1
Symbol - **H**
Atomic weight - 1.0079

- Alkali metals
- Alkaline earth metals
- Transition metals
- Rare earth metals
- Other metals
- Non-metals
- Halogens
- Noble (inert) gases

Period 1 2 3 4 5 6 7

# THE ATOM

Take a look at figure 2.3, the periodic table. Of these, only 25 are essential to biology, with carbon, hydrogen, nitrogen, and oxygen making up the vast majority of all living (or, I suppose, undead) matter. In fact, we're mostly going to concentrate on the first three rows. Let's take a closer look at one of these elements, nitrogen.

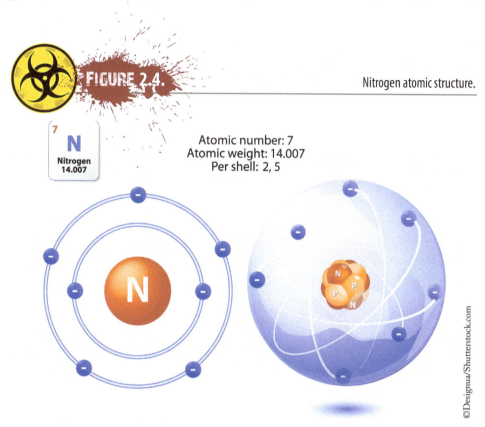

FIGURE 2.4.                                                    Nitrogen atomic structure.

7
N
Nitrogen
14.007

Atomic number: 7
Atomic weight: 14.007
Per shell: 2, 5

©Designua/Shutterstock.com

Here we see the structure of one nitrogen atom. An atom is the smallest structure that still retains the characteristics of the element. For this example, we can have as little as one atom of nitrogen, and it's still going to have all the properties of nitrogen. We can break it down further (and possibly have a very, very large explosion) but what we're left with isn't going to be nitrogen anymore. An atom is made up of positively charged protons along with neutrons (no charge) in its nucleus. The nucleus is surrounded by negatively charged electrons orbiting the nucleus in electron shells, not unlike planets revolving around the sun. As you can see, nitrogen has two electron shells. In fact, all elements on the second row along with nitrogen have two electron shells. So you can tell how many shells just by the row the atom is in. The innermost has two electrons, which is its maximum capacity. The second shell (and third shell, for those atoms that have one) holds a maximum of eight electrons. But no matter how many shells an atom has, only the outer shell will be important; it's called the valence shell. In fact, all of the inner shells are always full and stable. Nitrogen's valence shell has five of the eight spots full. More on why that is important in a bit.

There are a few other important pieces of information about the atom we can glean from the periodic table. First, each element has its own symbol, and in the case of nitrogen, it's "N." Also look at the number above

N. This is an element's atomic number, and it represents how many protons nitrogen has. We can see that nitrogen has seven protons. That is to say *all* nitrogen atoms in existence have seven protons. If one had one more or less proton, it simply wouldn't be nitrogen. The number below N is nitrogen's atomic mass, and it represents the number of protons and neutrons in the atom. You can see nitrogen has an atomic mass of 14.007. Now we've already established that it has 7 protons. But of course you can't have 7.007 neutrons. That's because unlike the number of protons, the number of neutrons can vary. So that means the *average* number of neutrons in a nitrogen is 7.007. The vast majority of atoms have seven; however a small percentage actually have eight. Atoms with a different number of neutrons are known as isotopes. For our purposes that's the last you'll hear about neutrons and isotopes, so just store it somewhere in case you ever get asked about them on *Jeopardy*. Initially an atom will have as many electrons as protons. We call these elemental atoms. Here's the problem. Atoms *hate* having an incomplete valence shell. They are highly unstable, volatile, perhaps dangerous, and overall poor conversationalists.

MOVIE SPOTLIGHT: *Zombeavers* (Rubin, 2014)
Causative agent—toxic chemical spill
Inexplicably transmissible for a chemical agent
It is suggested that the cause is the disease giardiasis, caused by the protozoan *Giardia intestinalis*, possibly mutated by the chemical
"We cannot turn against each other right now. That's exactly what the beavers would want!"

## BONDS AND MOLECULES

Let's take a look at two more atoms from the table.

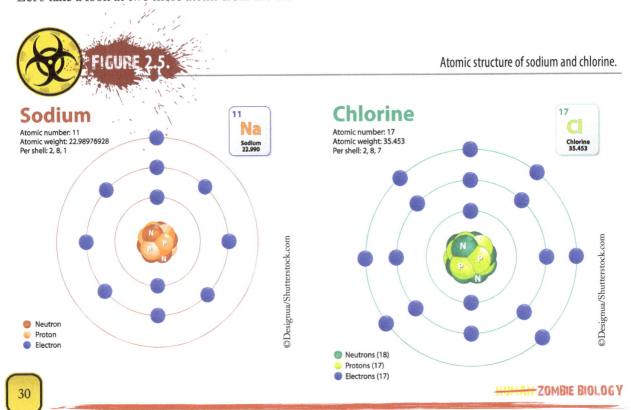

FIGURE 2.5.

Atomic structure of sodium and chlorine.

**Sodium**
Atomic number: 11
Atomic weight: 22.98976928
Per shell: 2, 8, 1

11
**Na**
Sodium
22.990

●  Neutron
●  Proton
●  Electron

©Designua/Shutterstock.com

**Chlorine**
Atomic number: 17
Atomic weight: 35.453
Per shell: 2, 8, 7

17
**Cl**
Chlorine
35.453

●  Neutrons (18)
●  Protons (17)
●  Electrons (17)

©Designua/Shutterstock.com

Sodium has one electron in a valence shell that holds eight, while chlorine has seven out of eight spots filled. So just how dangerous is elemental sodium? It can actually *explode* when it comes into contact with water. There's even usually enough water in the atmosphere to catch sodium on fire. So burning in air and exploding in water? Yeah you could say that's pretty volatile. And elemental chlorine? It's a poisonous gas that can kill any living thing on the planet, and was the first chemical warfare agent used in WWI. Again, not a happy little atom. So how do we fix this? Our friend the bond.

**FIGURE 2-6.**

Americans preparing for chemical attack, WWI France.

©Susan Law Cain/Shutterstock.com

The easiest way to do that is simply for one atom to give electrons to another. As stated earlier, chlorine needs just one electron to fill its outer shell. It can take one from sodium, which would lose the only electron in its outer shell. With an empty third shell (or more accurately, no more third shell), its full second shell is now its valence shell.

Now sodium has 11 positive protons and only 10 negative electrons, resulting in an atomic charge of +1. Chlorine in turn has 17 positive protons and 18 negative electrons, giving it an atomic charge of -1. These atoms that hold charges due to a difference in the number of protons versus electrons as a result of this swap are known as ions. They are now $Na^+$ and $Cl^-$, respectively.

The old adage that opposites attract holds true even on an atomic level, as the positive sodium ion is now attracted to the negative chlorine ion. In other words, $Na^+ + Cl^- \leftrightarrow NaCl$

And what happens to *explode in water* sodium and *WWI weapon* chlorine? Well they become NaCl, the chemical known as sodium chloride, or you might know it as table salt.

FIGURE 2-7.

Table salt.

©YuriyK/Shutterstock.com

These ions have formed an ionic bond, as their opposite charges are drawn to each other. This isn't the strongest bond there is, however, as if they split apart, both Na+ and Cl- would still have complete valence shells (and still be happy little atoms).

There are stronger bonds as well. Instead of one atom giving an electron to another, two atoms could share electrons instead. So once again, to the table!

FIGURE 2-8.

Oxygen atomic structure.

Atomic number: 8
Atomic weight: 15.999
Per shell: 2, 6

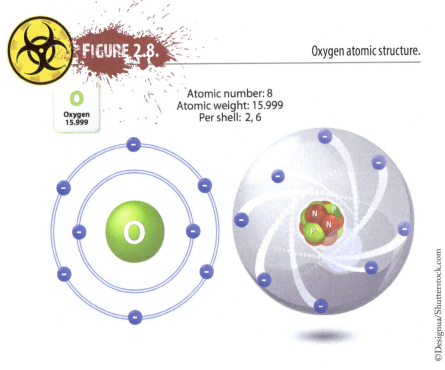

Oxygen
15.999

©Designua/Shutterstock.com

Fellow living breathing pre-zombies, oxygen is pretty damned important to us. You don't believe us? Try stopping for a few minutes. We'll find out why it's so important a little later in this book. Turns out what we breathe is actually two oxygen molecules bonded together, or $O_2$. As you can see from figure 2.8, oxygen has six of eight spots filled in its valence shell. Well if you have two oxygen molecules, you can't solve the problem by passing electrons. If one oxygen gave the two electrons the second needed, then it would be left with four in its valence shell and that's no help at all. So they each share two electrons with the other. Now four electrons are making a figure 8 around each nucleus, and at any one time each nucleus has the complete eight electrons orbiting it. This is called a covalent bond. This type of bond is much stronger than an ionic bond, because they need to stay bound to each other or become unstable. Specifically, since each oxygen is sharing *two* electrons, it's called a double covalent bonds. There are also single and triple covalent bonds. We show this by one, two, or three dashes, like so.

$$H_2, \quad \text{or} \quad H-H$$

$$O_2, \quad \text{or} \quad O=O$$

$$N_3, \quad \text{or} \quad N \equiv N$$

Now take $H_2O$, or water, for example. We know oxygen needs two electrons to complete its outer shell. In the case of water, it simply shares one with each hydrogen, as they each need one apiece to complete theirs. So we express water like this:

$$H_2O, \quad \text{or} \quad \begin{matrix} O-H \\ | \\ H \end{matrix}$$

Notice $H_2O$ is a bent, or polar molecule. This is in contrast to nonpolar $O_2$. If you remember the phospholipid bilayer of the cell membrane, it stops anything polar like $H_2O$ or ions like $Na^+$ and $Cl^-$ from coming through.

The molecules that make up all living (and undead) things on the planet are much larger molecules, all containing carbon. We call these molecules organic, as opposed to inorganic $H_2O$ or $O_2$. What's so special about carbon? Sigh... last time we promise, back to the periodic table. Notice it needs four electrons to fill its outer shell. In order to do that, it can form up to four separate covalent bonds. It can even form single, double, or triple bonds as well.

MOVIE SPOTLIGHT: *The Crazies* (Romero, 1973)
Causative agent—The military's newest biological weapon, a virus that causes victims to become violently insane.
A plane crash releases the virus into the town's water supply. A quarantine sets up a conflict between the military, the uninfected townspeople, and... *The Crazies.*
"Everything was going OK at first. People were willingly coming in with us. The US Army has medical facilities at the school. They're treating them at the school. But the ones with the virus... man, it's like they're crazy!"

FIGURE 2.9.

Examples of organic molecules.

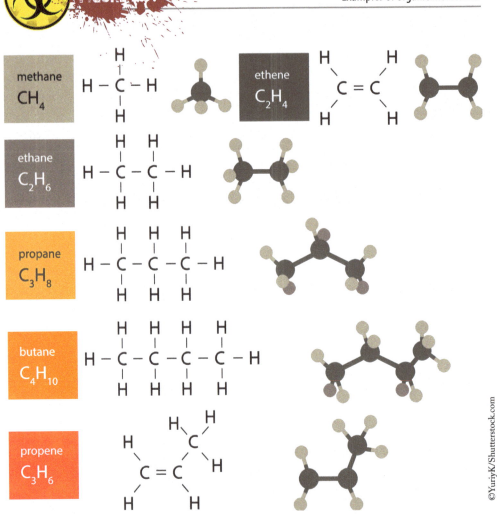

methane
$CH_4$

ethene
$C_2H_4$

ethane
$C_2H_6$

propane
$C_3H_8$

butane
$C_4H_{10}$

propene
$C_3H_6$

©YuriyK/Shutterstock.com

There's really no end to the forms that a chain of carbon can take. It can form chains of molecules thousands of atoms long. These large organic molecules are the building blocks of your cells. We call them macromolecules.

# CHARGE, HYDROPHOBICITY, AND SHAPE

Three characteristics of **molecules** determine the three-dimensional shapes of molecules and how they interact with each other. Many **macromolecules** (large molecules) are built as chains of smaller pieces. Often these smaller pieces have chemical characteristics, such as positive or negative charges (caused by having fewer or extra electrons compared to protons).

Negatively charged atoms and positively charged atoms may form larger molecules by sharing electrons, and lose their overall charge this way, but they may share those electrons unequally, and so still be "hungry" for an extra electron or proton on one end of the molecule. We call these molecules **polar molecules**, as they have a slightly more positive end ($\delta+$) and a slightly more negative end ($\delta-$). These polar molecules then like to form associations called **hydrogen bonds** with each other; this isn't a type of bond like a covalent bond where electrons are shared, but more of an association, the way two magnets are attracted to each other.

Charged and polar molecules mix well with water, so we call them hydrophilic.  Non-polar molecules do not mix well with water, so we call them **hydrophobic**. In the watery environment of a cell on earth, hydrophobic parts of molecules fold inwards, away from the water, while charged and hydrophilic parts move outwards to touch the water. This gives the strings of molecules their folded-up, three-dimensional shapes. These shapes are then stabilized by hydrogen bonds and other bonds between parts of the molecule.

Those 3D shapes then must fit together for interactions to take place. It works almost like puzzle pieces, with opposite charges attracting each other, similar charges repelling each other, and hydrophobic parts wanting to move inwards to be away from the water in cytoplasm. These puzzle pieces can then interact with other macromolecules. This is how enzymes interact with their substrates at active sites.

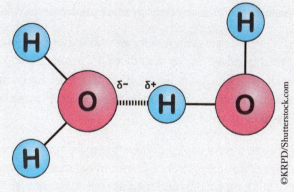

Water molecules forming a hydrogen bond (dotted) between the partially negative side (oxygen) of one molecule and the partially positive side (hydrogen) of the next molecule.

## PROTEIN STRUCTURE

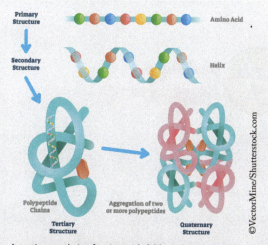

A continuous chain of amino acids folds up due to hydrophobic parts, hydrophilic parts, and hydrogen and disulfide bonds to form a 3D structure.

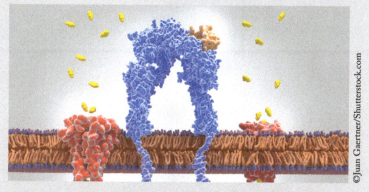

Insulin (orange proteins) interacting with its receptor on the cell surface (blue proteins) to open the channels (red protein channels) for glucose (yellow) to enter cells.

## Organic Macromolecules

There are four classifications of macromolecules. We'll summarize each of them.

Carbohydrates are carbon rings we call sugars (monosaccharides and disaccharides), or long chains of rings (polysaccharides). Sugars are one of the primary components of our diet and are used for energy. Starch is a long chain of glucose molecules that plants use for energy storage. Animals use a similar molecule, glycogen, for their primary energy storage in the liver. Plants also have a tough mesh polysaccharide called cellulose that makes up their cell wall.

Lipids are a structurally diverse group of macromolecules that have one thing in common: they are all hydrophobic, or repel water. The fat in our diet consists of molecules called triglycerides. The same molecules are in our own fat cells and used as energy storage. You may also remember the phospholipids that make up the cell membrane.

**FIGURE 2.10**

Molecular structure of sugars.

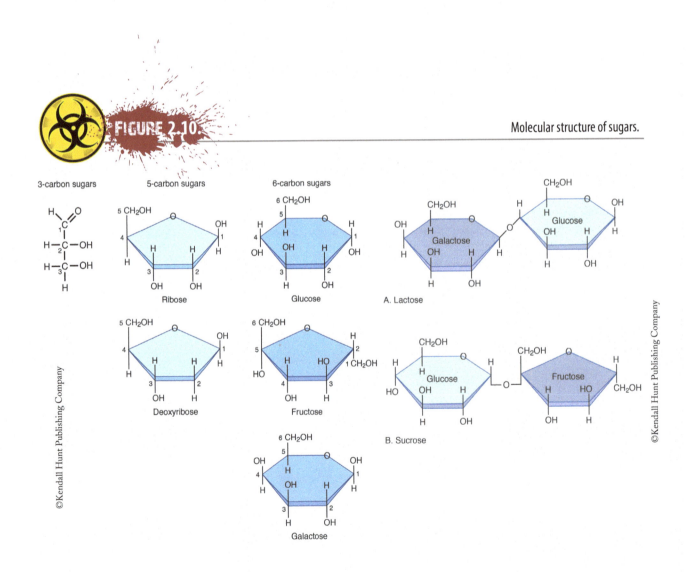

HUMAN ZOMBIE BIOLOGY

Proteins are massive organic molecules, thousands of carbons long. Many are structural, such as the actin and myosin proteins that make up your muscle. Many more are biological catalysts called enzymes, and they help perform just about every chemical reaction in a cell. More about them soon.

Finally, nucleic acids house your genetic information. Every physical or physiological characteristic an organism possesses can be found in its DNA and RNA. These are extremely complex molecules we'll learn more about in chapter 5.

So that's it for the basics of chemistry as succinctly as we can put it. Finally we can get back to zombies.

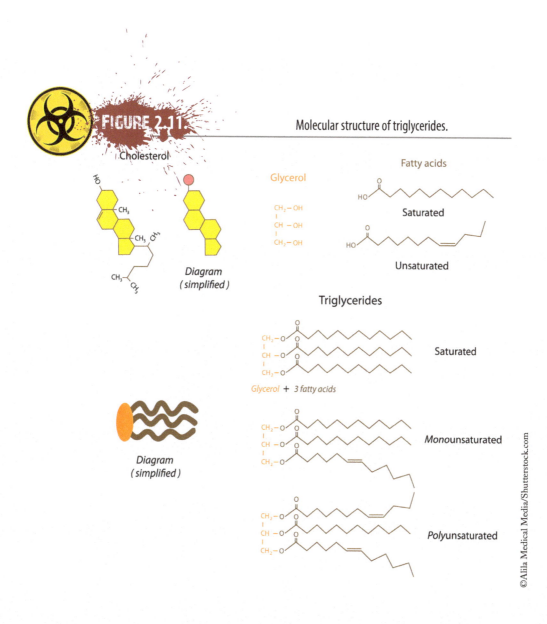

FIGURE 2.11 — Molecular structure of triglycerides.

 FIGURE 2-12.

The protein rhodopsin, essential for vision.

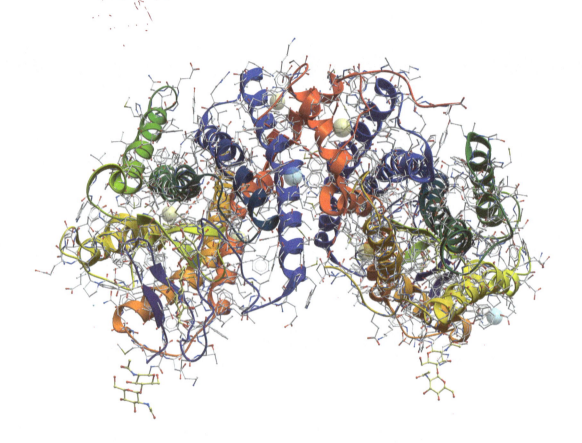

1. The sodium atom contains 11 electrons, 11 protons, and 12 neutrons. What is the atomic mass of sodium?
   a. 0
   b. 11
   c. 22
   d. 23
   e. 34

2. The atomic number of an atom is
   a. the number of protons in the atom.
   b. the number of electrons in the atom.
   c. the number of neutrons in the atom.
   d. the number of protons, electrons, and neutrons in the atom.
   e. the net electrical charge of the atom.

3. What is the fundamental difference between covalent and ionic bonding?
   a. In a covalent bond, the partners share a pair of electrons; in an ionic bond, one partner captures an electron from the other.
   b. In covalent bonding, both partners end up with filled outer electron shells; in ionic bonding, one partner does and the other does not.
   c. Covalent bonding involves only the outer electron shell; ionic bonding also involves the next inner electron shell.
   d. Covalent bonds form between atoms of the same element; ionic bonds, between atoms of different elements.

4. The nucleus of an atom contains
   a. protons and neutrons.
   b. protons and electrons.
   c. only neutrons.
   d. only protons.
   e. only electrons.

5. Which of the following, if any, is *not* a characteristic of all living organisms?
   a. Respond to the environment.
   b. Develop and maintain complex organization.
   c. Take in and use energy.
   d. Maintain homeostasis.
   e. All of the choices are characteristic of all living organisms.

6. Which sentence about electrons is false?
   a. Electrons travel around the nucleus in electron shells.
   b. Electrons carry a negative charge.
   c. Electrons are located in the nucleus along with neutrons.
   d. Elemental atoms have the same number of electrons as protons.

7. _____ are the weakest bonds, not strong enough to hold atoms together to form molecules but are strong enough to form bridges between molecules.
   a. Ionic bonds
   b. Covalent bonds
   c. Polar covalent bonds
   d. Hydrogen bonds
   e. Common bonds

8. Organic compounds
   a. always contain nitrogen.
   b. are synthesized by only animal cells.
   c. always contain carbon.
   d. can only be synthesized in a laboratory.
   e. always contain oxygen.

| WORDSTEMS | |
|---|---|
| -cule | little |
| -elle | diminutive |
| eu- | well, true |
| homeo- | same |
| hydro- | water |
| karyo- | nucleus |
| lipo- | fat |
| macro- | large |
| meta- | change; behind |
| -philic | loving |
| -phobic | fear; hate |
| pro- | first |
| -stasis | standing; stabilize |
| trans- | strength or power (of combining with other atoms) |

# CHAPTER 3

## ZOMBIE METABOLISM

**(OR... WHEN EATING HUMANS AVOID A HIGH FAT, ALL COUCH POTATO DIET)**

In popular culture, zombies represent an "other"; a non-human creature to be confronted and overcome. Most of the stories, comic books, movies, and video games about zombies continue with an early zombie characteristic... that zombies are mindless, homicidal predators hungry for human flesh (or, in some stories brains, and if you prefer a taste of the theatrical, Braaaaaiiiiinnnnsss).

In fact, zombies characterized as the dead who eat the living are recorded in one of the first examples of writing. In The Epic of Gilgamesh, written circa 2000 BCE, the goddess Ishtar calls for revenge:

"I will smash in the gates of the netherworld; I will set up the ruler of the great below, and I will make the dead rise, and they will devour the living, and the dead will increase beyond the number of the living" (George, 2003).

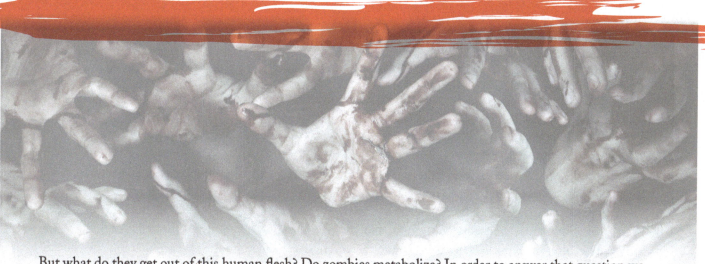

But what do they get out of this human flesh? Do zombies metabolize? In order to answer that question we need to delve deeper into what exactly metabolism is.

FIGURE 3.1.

Enzymatic reaction.

## The Lock and Key Mechanism

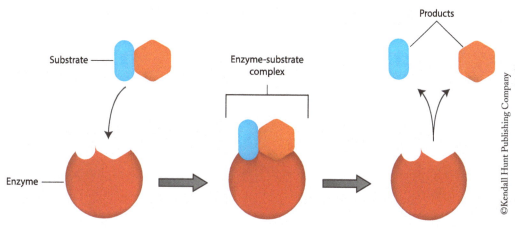

©Kendall Hunt Publishing Company

## ENZYMES

Picture this scenario. You find yourself nodding off during your biology lecture and your professor slams a book down on your desk to wake you. It's clear from the look your professor is giving you (as well as the snickers from your classmates) that he or she is expecting an answer and you missed the question. If you ever find yourself in this dire situation, tell the professor "enzymes" as you have a 1 in 4 chance of being right.

As stated earlier, life revolves around metabolism, the process of converting energy from one form to another. Cells use enzymes to build and break bonds between atoms. Bonds take energy to form, and release

energy when they are broken. Using enzymes means that less energy is needed to break or build bonds, and therefore a little energy can go a long way!

Enzymes work by having just the right shape to bring atoms in close proximity to one another, and providing a little extra tension on bonds (to break them) or a little extra push (to build them).

Some enzymes synthesize (build larger molecules from smaller ones). Since bonds are being formed, these metabolic reactions require energy input and are called anabolic reactions. All this means that cells are effectively storing energy in larger molecules. Other organisms can turn around and steal this energy by eating those molecules, like sugar and starch and complex proteins. Then, when the cells break down the larger molecules, small packets of energy are released. These are called catabolic reactions. But the cell still needs to get those packets of energy to where it's needed… kind of the way we use batteries. More on that in a second.

There is a third, very important type of metabolic reaction called a redox reaction. Redox is short for *red*uction and *oxi*dation reactions. These reactions neither require nor release energy; instead they involve the transfer of electrons from one molecule to another. Molecules that lose an electron are said to be oxidized, while those that gain electrons are said to be reduced.

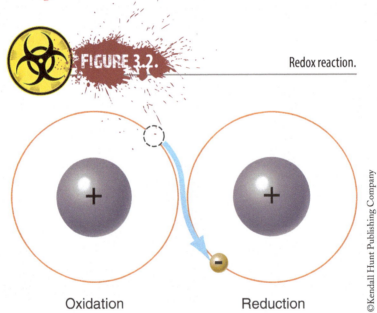

**FIGURE 3.2.**    Redox reaction.

Oxidation        Reduction

## ATP, YOUR BIOLOGICAL BATTERY

One example of an energy carrier molecule that our body uses is called adenosine triphosphate (ATP). Figure 3.3 shows the structure of ATP. Notice the three phosphate (P) groups at the end. Each of the bonds between the phosphates in this molecule is a high-energy bond, which can be broken and the energy transferred to enzymes and cell parts that need it.

FIGURE 3.3.

Adenosine triphosphate molecular structure.

NH₂

N

N

N

N

Adenine

O

O

O

O

⁻O–P–O–P–O–P–O–CH₂

O⁻

O⁻

O⁻

5

4

3

2

1

H

H

H

H

OH

OH

Triphosphate group

5-carbon sugar

©Kendall Hunt Publishing Company

When your body needs energy, it breaks the first phosphate group to release the energy stored there. Hence adenosine *tri*phosphate becomes adenosine *di*phosphate (ADP). Later, your body can reuse the ADP by replacing the phosphate group, turning it back into ATP. So, one way to think about it is a metaphor… where ATP is a rechargeable battery that can be used by enzymes that break (catabolic enzymes) and enzymes that build (anabolic enzymes).

FIGURE 3.4.

Your cells (the drill) use ATP (the battery) until it runs out of juice… then your body recharges ADP making ATP again

## HYDROLYSIS ENZYME

## ATP

## SYNTHESIS ENZYME

Source: Stephanie Fischer

BATTERY-POWERED SAW

 INTERCHANGEABLE BATTERY

BATTERY-POWERED SCREWDRIVER

## RECHARGING YOUR BATTERIES

So remember, the next question is, how do the batteries (ATP) get recharged?

Plants and algae (and certain bacteria) undergo a process called photosynthesis are able to charge ATP from sunlight. Animals, along with protozoans, fungi, and all disease-causing bacteria, are able to charge ATP by breaking down large organic molecules such as sugars and proteins and harvesting the energy in the bonds. You and I need oxygen and undergo an aerobic (with air) process called cellular respiration. Simpler organisms that don't use oxygen undergo an anaerobic (without air) process called fermentation.

## CELLULAR RESPIRATION

Cellular respiration is the process by which we break down our food (glucose) to harvest the energy in the bonds. In summation:

$$C_6H_{12}O_6 + 6O_2 \longrightarrow 6CO_2 + 6H_2O + energy$$

Or... glucose and oxygen get changed to carbon dioxide. We break the bonds of the large glucose molecule and harvest enough energy to turn 38 ADP back into 38 ATP. In other words, recharge our biological batteries. It should be noted that the equation above is simply a *summary* of cellular respiration, and in actuality it's metabolic reactions one after the other. Each of these steps needs a separate enzyme to proceed as well, so dozens of enzymes are also involved in the process.

Most of our energy actually comes from stripping electrons away from glucose (oxidizing it) and reducing complex electron carriers such as $NAD^+$ into NADH (reduced form). These reduced molecules with the electrons stolen from glucose then proceed to the electron transport chain located in the innermost layer of the mitochondrion. This inner mitochondrial membrane (IMM) is a phospholipid bilayer similar to the cell membrane.

The electron transport chain uses a system of enzymes to pass electrons from one to the next. Each time an electron is passed, the enzyme uses that energy to pump a proton ($H^+$) to the other side of the membrane. This creates a gradient. You can think of it a little like carrying water up a hill, and then once the water is up the hill, you can use the movement of the water down the hill to turn a wheel. Because molecules find gradients energetically unfavorable, they will diffuse (move down the gradient) to reach equilibrium.

Once a gradient of protons has built up on one side of the membrane, the protons want to return to equilibrium. Remember, anything with a charge can't pass through the phospholipid bilayer, so the protons must come through a specialized enzyme and channel called ATP synthase. ATP synthase is a wheel-like enzyme that moves like a stadium turnstile. As ATP synthase turns, it "charges" the ATP battery by attaching a phosphate to ADP.

**FIGURE 3.5.**

Electron transport chain generating ATP in the inner mitochondrial membrane.

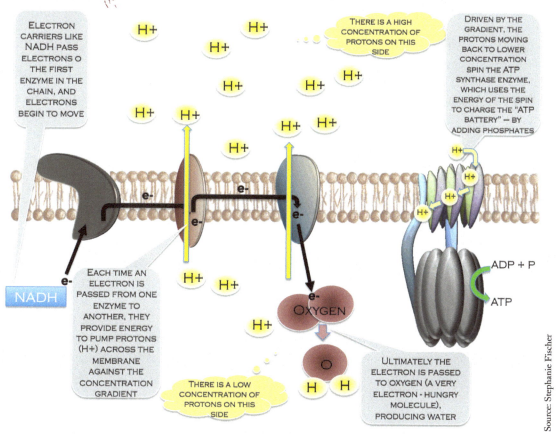

ELECTRON CARRIERS LIKE NADH PASS ELECTRONS O THE FIRST ENZYME IN THE CHAIN, AND ELECTRONS BEGIN TO MOVE

THERE IS A HIGH CONCENTRATION OF PROTONS ON THIS SIDE

DRIVEN BY THE GRADIENT, THE PROTONS MOVING BACK TO LOWER CONCENTRATION SPIN THE ATP SYNTHASE ENZYME, WHICH USES THE ENERGY OF THE SPIN TO CHARGE THE "ATP BATTERY" — BY ADDING PHOSPHATES

EACH TIME AN ELECTRON IS PASSED FROM ONE ENZYME TO ANOTHER, THEY PROVIDE ENERGY TO PUMP PROTONS (H+) ACROSS THE MEMBRANE AGAINST THE CONCENTRATION GRADIENT

THERE IS A LOW CONCENTRATION OF PROTONS ON THIS SIDE

ULTIMATELY THE ELECTRON IS PASSED TO OXYGEN (A VERY ELECTRON-HUNGRY MOLECULE), PRODUCING WATER

NADH

OXYGEN

ADP + P

ATP

Source: Stephanie Fischer

## OXYGEN'S ROLE IN MAKING ATP

So the ET chain and ATP synthase convert our ADP to ATP. That's all well and good, but we still have a few things to worry about. First, as the protons come back in through ATP synthesis, the gradient weakens. No gradient = no ATP production. So we've got to find a way to get rid of some of these protons.

But we have a bigger issue. Remember those electrons we gave to the ET chain? Well, a lone electron can cause all sorts of damage inside your cell. Blowing through membranes, tearing apart molecules, it's like a bullet ricocheting from one target to another. So we've also got to get rid of electrons.

Enter oxygen. It can bind with the loose protons and electrons, fixing both problems at the same time.

$$\tfrac{1}{2}O_2 + 2H^+ + 2e^- \longrightarrow H_2O$$

As oxygen binds to the loose protons and electrons it's converted to metabolic water. Oxygen is the key player in how we change what we eat into energy. It really is kind of a control-burn happening in the mitochondria of our cells, with oxygen being one part of the fuel, just like a fire.

So, if our cells lack oxygen… our cells can't make more ATP… and our cells die. If we lack oxygen to critical organs, those critical organs can't make more ATP… and we die.

So that begs the question… our reanimated corpses were originally corpses specifically because they lacked ATP—our (slightly oversimplified) ultimate cause of death in the zombie apocalypse.

Then they rose up and became the "walking dead." How are they moving? Now let's think about that for a minute….

So how does it work for us still among the living?

## ATP'S ROLE IN MUSCLE CONTRACTION

ATP plays a role in three critical parts of muscle contraction: the neuron signal to start the contraction, the release of the muscle fibers after contraction, and the return of **calcium ions** into storage after muscle contraction.

Skeletal muscle is composed of specialized cells full of two protein fibers that are able to pull against each other, in a kind of ratchet or "hand over hand" type of movement; its useful to think of sailors hauling on rope, hand-over-hand, to think about it. The **myosin** protein has little "heads" that attach into spots on the **actin** protein and pull against it.

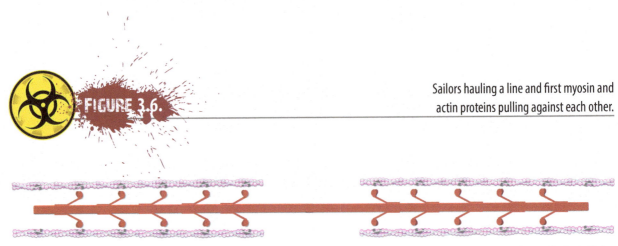

**FIGURE 3.6.**

Sailors hauling a line and first myosin and actin proteins pulling against each other.

Source: Stephanie Fischer

Muscle cells have lots of these myosin protein fibers and actin protein fibers, but also lots of mitochondria to make the needed ATP, and also a special structure called the **sarcoplasmic** reticulum that stores calcium ions.

FIGURE 3.7.

Muscle cells contain myosin and actin fibers, as well as mitochondria to make ATP and the sarcoplasmic reticulum, a storage network for calcium ions.

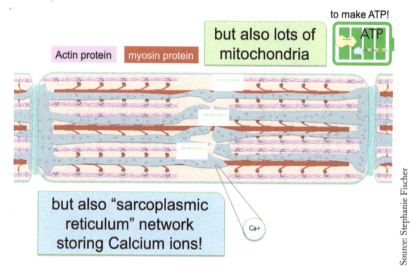

Source: Stephanie Fischer

The actin protein has binding sites that perfectly fit the myosin head; but at rest, those binding sites are covered by another protein doublet named **troponin** and **tropomyosin**. Tropomyosin only moves away from those binding sites when its partner troponin is bound to calcium ions.

We will talk more about neuron signals in the Brains chapter, but briefly, ATP is required as an energy source to allow neurons to signal to muscle cells that the time for a contraction is NOW. At that point, the neuron signal causes the calcium ions to be released (through a **voltage-gated ion channel**) from the sarcoplasmic reticulum storage, and the calcium ions move troponin and tropomyosin away from the binding sites on the actin.

The myosin heads can then do what they energetically want to do, which is bind to the actin and pull against the actin, shortening the cell overall by a miniscule amount. It is the collective action of all of the microscopic proteins pulling, and then releasing and pulling again and again, that creates movement.

The energy that is required, though, is critical for the "again and again" part of that sentence. ATP is required for the release of myosin heads from actin, so that after the initial tiny pull, the heads can release and re-bind to the next slot and pull again. ATP is not required for the first ratchet movement, but is required for any further ratcheting to continue to shorten the muscle cell. And ATP is required for the ultimate release and then relaxation of the muscle, and to pump the calcium back into storage away from the troponin/tropomyosin, also required for the relaxation of the muscle.

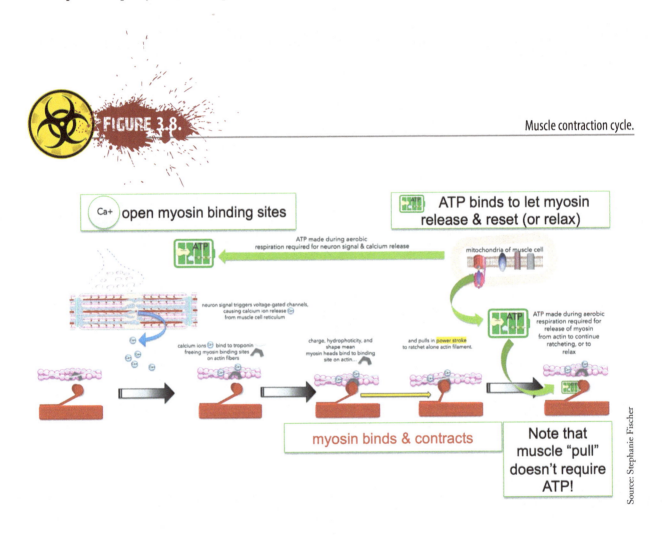

FIGURE 3.8.                                                                            Muscle contraction cycle.

Ca+ open myosin binding sites

ATP binds to let myosin release & reset (or relax)

ATP made during aerobic respiration required for neuron signal & calcium release

mitochondria of muscle cell

ATP

neuron signal triggers voltage-gated channels, causing calcium ion release from muscle cell reticulum

calcium ions bind to troponin freeing myosin binding sites on actin fibers

charge, hydrophoticity, and shape mean myosin heads bind to binding site on actin...

and pulls in power stroke to ratchet alone actin filament.

ATP made during aerobic respiration required for release of myosin from actin to continue ratcheting, or to relax

myosin binds & contracts

Note that muscle "pull" doesn't require ATP!

Source: Stephanie Fischer

Therefore, ATP is required for the initial neuron signal that causes the calcium release, but it is also required for repeated ratcheting and for the relaxation of the muscle.

Rigor mortis, then, the contraction-without-relaxation that occurs in organisms that are dead and no longer making ATP, makes sense. About 13 hours after death, calcium is released in an uncontrolled way from the sarcoplasmic storage as decomposition causes the channels to break down, and that triggers a cycle of muscle contraction. But no additional ATP is being made after death (no oxygen, no ATP production; remember aerobic respiration a few pages ago). So then no muscle relaxation can occur, and the rigor (or stiffness) continues until the actin and myosin begin to break down themselves, resulting in the final relaxation of death ("secondary flaccidity") between 48 and 80 hours later.

## OTHER SYSTEMS FOR MAKING ATP

So discounting movies like *28 Days Later*, it's clear your average zombie doesn't need oxygen in the slightest. This begs the question, how are they recharging their ATP without breathing in oxygen? There are other methods of recharging ATP without oxygen. As stated earlier, more primitive organisms like fungal yeast and many bacteria use a simpler method called fermentation that doesn't need oxygen. It's a quick, easy method of breaking sugar in two and converting two ADP to ATP. The waste product of fermentation is either an acid or an alcohol. It's the process that makes our bread rise and gives us beer, wine, cheese, and yogurt.

Turns out we can do something very similar. Because ATP is so critical to the function of an organism, our muscles have two other ways to make ATP even without oxygen. The phosphagen system and the glycogen–lactic acid system can generate ATP for very short periods of time (measured in seconds or a few minutes), which is very important for short bursts of energy for our muscles, but not enough to keep us alive if our critical organs are oxygen starved. When you are performing intense exercise like sprinting or lifting heavy weights your oxygen will be the first component of cellular respiration to run out. This anaerobic workout triggers the alternate systems.

So… #1 if muscles can't work without ATP and #2 zombies don't breathe (and #3 zombies don't turn into photosynthesizing plants) then the most feasible possibility is that these zombies may be fermenting. But it would have to be some extra-efficient paranormal fermentation, to be able to produce enough ATP to move around a whole human body for any length of time. But if they're producing ATP by paranormal fermentation… are they truly dead?

MOVIE SPOTLIGHT: *Re-animator* (Gordon, 1985)
Causative agent—chemical reanimating agent
Based on the short story *Herbert West – Reanimator* by H.P. Lovecraft (1922)
Theatrical cut given an X rating by MPAA
"Who's going to believe a talking head? Get a job in a sideshow."

## ATP, LIFE AND DEATH

The argument follows this way:

In (oversimplified but real) terms, an organism is dead when it no longer metabolizes (produces or uses ATP) or reproduces its cells.

# DOES ELECTRICITY REGENERATE ATP?

Mary Shelley wrote what is arguably the first science fiction novel, *Frankenstein, or the Modern Prometheus*, in 1818. As we will read about in the chapter about the brain, people at the time were beginning to scratch the surface of the mechanisms, and dangers, of electricity's interaction with living things. Some "experts" at the time (who were often just sideshow entertainers) would ask people to pay to observe them run an electrical current through a dead frog, for example, and watch the frog "come back to life", giving rise to a lot of mythology about reanimated corpses. In reality, the electricity was opening those **voltage gated ion channels** in neurons and muscle cells, causing one big rush of **calcium ions**, and letting the muscles perform one big contraction. But the animals were truly dead, with no generation of ATP.

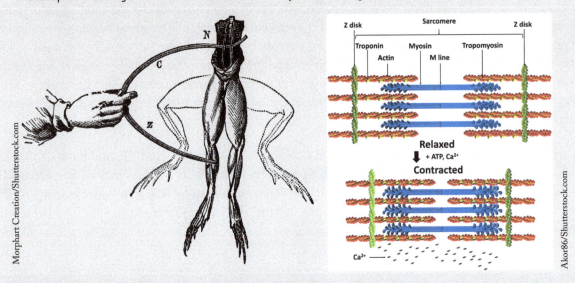

With no ATP being produced, there was no continued contraction and no relaxation of the muscles, so no actual controlled movement or true brain activity.

One twitch is all you get with that kind of electrical manipulation. While Shelley brought up important questions in the novel about the meaning of life, the responsibility of scientists and doctors, and even the responsibilities of parents, electricity cannot reanimate a corpse or be a source of zombies.

If our outbreak zombies were in fact truly dead, no ATP is being produced in their cells (the ultimate definition of dead). If they are truly dead, their cells are not doing aerobic respiration to produce ATP, nor the theoretical paranormal fermentation we talked about earlier.

With no ATP, the muscle fibers can't relax to restart muscle contractions.

With no ATP, the neurons can't fire to trigger the muscle contractions.

With no ATP, the neurons in the brain can't communicate.

So, how the heck are the zombies moving?

On the other hand, if they ARE producing ATP (metabolizing) and then using the ATP to move, then technically... they aren't dead. Which means, they aren't zombies. But that's no fun at all, is it?

# CHAPTER 3

## QUESTIONS/WORDSTEMS

1. Which of these DO NOT slow down enzymatic activity rates?
   a. Decreased substrate concentration
   b. Denatured enzyme
   c. Presence of an inhibitor
   d. Decreased enzyme concentration
   e. Increased substrate concentration

2. Write the equation for cellular respiration. How much ATP can be produced when one molecule of glucose is completely broken down?

3. What is the role of oxygen in cellular respiration?

4. Give three differences between cellular respiration and fermentation.

| WORDSTEMS | |
|---|---|
| -aerobic | with oxygen |
| an- | without |
| ana- | upward |
| -ase | enzyme |
| cata- | downward |
| -lysis | to break |
| photo- | light |
| -spire | breathe |
| synth- | to build |
| -zyme | yeast |

# CHAPTER 3

## WORKSHEET

## HOW ENZYMES WORK (GROUP PROJECT)

Adapted with permission from (Green, Clark, Mathis, Barkhurst, & Mansfield, 2015)

### OBJECTIVES

1. To introduce the effect of enzymes on substrate
2. To study the effects of alteration of enzyme activity

### INTRODUCTION

Enzymes work as organic catalysts in living cells to control the rates of chemical reactions. Many enzymes are important in catabolic reactions in the overall metabolism of a cell. During metabolism, the enzyme attaches its active site to a substrate molecule and promotes the breaking of chemical bonds of the substrate. The products of this action are two smaller molecules and the release of energy. The enzyme releases the products and will attach to another substrate molecule and continue breaking bonds. The enzyme will continue breaking down substrate as long as the substrate is available and as long as the enzyme's active site is open and properly shaped. If an enzyme becomes denatured (loses the proper shape of its active site) or inhibited (the active site becomes blocked by substances other than the substrate), then the enzyme will be unable to continue to process the substrate.

### PROCEDURE

### MATERIALS

- 200 toothpicks
- 15 dissecting pins
- Masking tape
- Tennis ball
- Bowl

From *General Biology Laboratory Manual*, 2/E by Christopher F. Green, Krista L. Clark, Karen Mathis, Sue Barkhurst and Jennifer Mansfield. Copyright © 2016 by Kendall Hunt Publishing Company. Reprinted by permission.

- Stopwatch
- Blindfold
- Golf ball

Working in groups of three, one partner will be playing the role of an enzyme called "toothpickase" which has as its substrate, the substance toothpick. The action of "toothpickase" is to aid in breaking the bonds of toothpick, creating two smaller but equally sized substances (pieces). The second partner will count the number of toothpicks broken during each 10-second interval and record data. The third partner will use the stopwatch to indicate when each time interval is over so that the second partner can record the data properly. During this activity, you will determine the rate of reaction for "toothpickase" under four different conditions: an adequate supply of substrate, a limited supply of substrate mixed in with other substances that mimic the substrate, an enzyme that is partially denatured, and an enzyme that is inhibited.

Tape the "toothpickase's" second ring and pinky fingers to his or her palm. Only the thumb, index, and middle fingers are the active site for "toothpickase."

## ADEQUATE SUPPLY OF SUBSTRATE

- Place 40 toothpicks in small bowl. Designate one partner as "toothpickase" and blindfold that person. "Toothpickase" may only use one hand for this activity.
- The partner with the stopwatch will start the timer. "Toothpickase" will pick up toothpicks and break (metabolize) them in half, dropping broken portions back into the bowl. The second person will count the number of toothpicks broken in half during 10-second intervals and record. The third person will indicate when each 10-second interval is finished so that data can be recorded in the correct spot. The timer will run continuously for 2 minutes (therefore there will be a dozen 10-second intervals recorded). Calculate the cumulative number of toothpicks broken.
- Remove the broken toothpicks and throw them in the trash. Add a new supply of toothpicks to the bowl.

## LIMITED SUPPLY OF SUBSTRATE

- Instead of the bowl spread the 40 toothpicks around a baking pan.
- Repeat the process of picking up, breaking, and recording for another 2 minutes.
- Remove broken toothpicks and throw them away. Restock bowl with toothpicks.

## INHIBITED ENZYME

- Place 20 bobby pins in the bowl along with the 40 toothpicks.
- The bobby pins represent an inhibitor that blocks the active site of "toothpickase." Whenever "toothpickase" picks up a bobby pin instead of a toothpick, have them replace it back into the bowl.
- Remove broken toothpicks and throw them away. Restock bowl with toothpicks.
- Repeat the process of picking up, breaking, and recording for another 2 minutes.

## Denatured Enzyme

- You'll be given tape, a golf ball, and a tennis ball. Use these (and anything else you can think of) to change the shape of "toothpickase." Tape the fingers together, tape one of the balls to the front of your hand, whatever you want.
- This will mimic "toothpickase" being denatured.
- Repeat the process of picking up, breaking, and recording for another 2 minutes.
- Remove broken toothpicks and throw them away. Return all supplies to the designated area. Restock bowl with toothpicks.

## Results/Questions

| Time (seconds) | Metoabolism (number of toothpics broken) | | | | | | | |
| --- | --- | --- | --- | --- | --- | --- | --- | --- |
| | Ideal | | Limited Substrate | | Inhibited Enzyme | | Denatured Enzyme | |
| | Number | Cumulative | Number | Cumulative | Number | Cumulative | Number | Cumulative |
| 10 | | | | | | | | |
| 20 | | | | | | | | |
| 30 | | | | | | | | |
| 40 | | | | | | | | |
| 50 | | | | | | | | |
| 60 | | | | | | | | |
| 70 | | | | | | | | |
| 80 | | | | | | | | |
| 90 | | | | | | | | |
| 100 | | | | | | | | |
| 110 | | | | | | | | |
| 120 | | | | | | | | |

- When given limited substrate what happened to the rate at which "toothpickase" was able to work during the 2 minutes of activity compared to the ideal?

- What happened to the rate at which a denatured "toothpickase" was able to work during the 2 minutes of activity compared to the ideal?

- What happened to the rate at which inhibited "toothpickase" was able to work during the 2 minutes of activity compared to the ideal?

- Using MS Excel, create a scatter graph (with line) to show the metabolic rate for all four conditions. Be sure to include all necessary components (see worksheet 1). Time is the independent variable and is plotted on the X axis. The *cumulative* number of toothpicks metabolized is the dependent variable and is plotted of the Y axis. Paste the graph in the space below.

# SECTION 2

## How does one become a zombie?

# CHAPTER 4

## THE VOODOO CURSE

### (OR... "LET'S JUST AGREE TO AVOID HAITI THIS SPRING BREAK.")

Courtesy of Emily Adele

In order to answer this section's fundamental question of what makes a zombie, we should go back to where it all started. So pack your bug spray and SPF because we're traveling to the tropical paradise known as... Haiti. Sorry, I just checked our class funds... let's stay here and just read about it.

## HAITI AND THE BIRTH OF ZOMBIE LORE

The word "zombie" originated from Haitian rural mythology. Formal Haitian voodoo originated from the West African practice of *vodou*. From there it immigrated with the African slave trade to Haiti. In common rural folklore, a zombie was created by necromancy; a zombie was a corpse reanimated by a *bokor*, a vodou sorcerer or witch. This is opposed to the *houngan* or *mambo*, priests and priestesses of the formal voodoo religion.

**MOVIE SPOTLIGHT:** *White Zombie* (Halperin, 1932)
Causative agent—potion from evil zombie master
Based on the book *Magic Island* by William Seabrook, 1929
Widely accepted as the first zombie movie
"Your driver believed he saw dead men... walking."

Zombies were not part of formal Haitian voodoo, yet were still one of several widely held beliefs. Article of the Haitian Penal code 246, written in 1864, appears to give credence, defining murder to include zombieism (Organization of American States, 2013).

Art. 246.- Is qualified poisoning, all attempts on the life of a person, by the effect of substances that can cause death more or less promptly, however these substances have been used or administered, and whatever were the suites.- C. pén. 240, 247, 262, 263, 334, 372.

Also qualified attempt on the life of a person, by poisoning, the use made against it without substances that cause death, have produced a more or less prolonged lethargy, however these substances have been employed and whatever were the suites.

*If, following this lethargic state, the person was buried, the attack will be qualified* assassinat.- C. pén. 241 and following. Thus mod. Act 27 Oct. 1864.

In 1929, explorer and author (and by all accounts, complete nutjob) William Seabrook wrote of his trip to Haiti and his encounter with voodoo and zombies (Seabrook, 1929). The text and accompanying artwork by Alexander King would first introduce the concept of zombies to the American public and greatly influence popular culture. The first movie to feature zombies was 1932's *White Zombie*, a loose adaptation of Seabrook's book starring horror icon Bela Lugosi (Halperin, 1932).

**FIGURE 4.1.**

*White Zombie* theatrical movie trailer, 1932.

Halperin 1932

In 1985, ethnobotanist (yeah, that's a real career) Wade Davis published *The Serpent & the Rainbow*, documenting his travels to Haiti and witnessing the process of making zombies (Davis, 1985). Davis contended that "zombification" was achieved by certain chemicals such as tetrodotoxin (a deadly toxin found in puffer fish) and datura extract (a hallucinogenic plant). According to Davis, this combination of poisons slowed bodily functions to the point that the victim would be considered dead, even to a doctor.

In the book, Davis told the story of Clairvius Narcisse, a man declared dead by two doctors in Haiti and buried in 1962. Despite this apparent setback, Davis stated that he was found wandering around 18 years later. He had reportedly been given tetrodotoxin and other chemicals to zombify him where he worked on a sugar plantation as a "zombie" slave for two years.

Now… before you get all excited about the prospect of real-life zombies, it should be noted that several scientific articles refute Davis's claim, while analysis of samples Davis brought back showed little or no tetrodotoxin. Some actually accused Davis of all-out fraud.

Regardless, the question remains. Is it possible for a chemical to cause zombie-like symptoms? Can a chemical actually bring people back from the dead? Well if at all possible, these chemicals need to affect the organ that makes you… you. We'll need to delve into the inner workings of your brain and nervous system.

## THE BRAIN

Let's start with the most complicated, least understood organ in your body, the brain. Figure 4.2 shows the components of the brain.

FIGURE 4.2.

Diagram of the human *Braaaaiiiinnnn.*

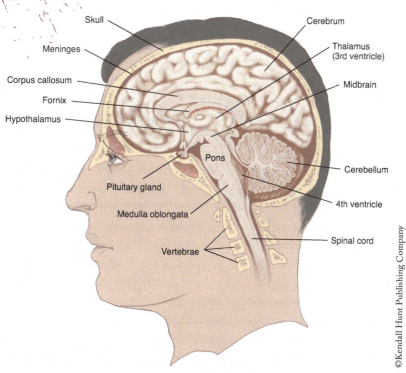

©Kendall Hunt Publishing Company

## THE CEREBRUM

The cerebrum is both the largest and most complex portion of your brain. In figure 4.3, you see it sectioned off into the frontal, temporal, parietal, and occipital lobes. You can correlate intelligence of species of animals based on the amount of cerebral cortex (grey matter). In primitive fish such as lampreys it is a simple structure that receives olfaction signals (smell). The cortex increases in size and complexity through fish, amphibians, reptiles, and birds. As higher mammals we enjoy the highest percent of cerebral cortex per volume (hence the shake weight and Macarena!).

The functions of the cerebrum are numerous and diverse. They include but are not limited to:

- Memory
- Learning
- Language and communication
- Senses (sense of sight, smell, hearing, taste, touch)
- Voluntary movement

One of the first things scientists and doctors noticed when they began to work with human brains is that the cerebrum has distinct areas in its structure. We have since created rough maps of the cerebral cortex, and figured out that certain types of functions are often localized together. The way these were really figured out was by carefully mapping out functions lost when patients had some kind of damage to the brain (stroke, seizure, or injury). When specific areas of the cortex were damaged, consistent effects were seen, so neurologists and neuroscientists were able to draw a map.

Most zombie stories focus on the idea that zombies are "mindless"; that they are simply trying to kill victims, or trying to dine on human brains, because they have lost the ability to reason. Some stories hypothesize some kind of microbe that effectively "causes a lobotomy". A lobotomy was used when we had minimal understanding of the brain, as a way to relieve symptoms of some types of severe psychosis. Popular culture often says that a lobotomy was a complete removal of the frontal lobe, which was actually not usually the case. It was less often a removal of the frontal lobe (which is the center of personality, control of behavior, planning, and execution), but it was more often a localized destruction of a particular part of the frontal lobe. Sometimes, specific connections were severed between areas of the frontal lobe and the rest of the brain. It was used in an effort to control severe depression, schizophrenia, mania, and other severe forms of psychosis. Complications from the procedure ranged from problems controlling basic functions (temperature, heart rate, breathing rate) due to severing the connections between the frontal lobe and serotonin nuclei in the

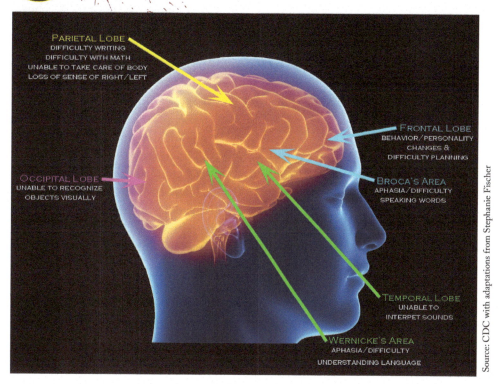

PARIETAL LOBE
DIFFICULTY WRITING
DIFFICULTY WITH MATH
UNABLE TO TAKE CARE OF BODY
LOSS OF SENSE OF RIGHT/LEFT

FRONTAL LOBE
BEHAVIOR/PERSONALITY
CHANGES &
DIFFICULTY PLANNING

OCCIPITAL LOBE
UNABLE TO RECOGNIZE
OBJECTS VISUALLY

BROCA'S AREA
APHASIA/DIFFICULTY
SPEAKING WORDS

TEMPORAL LOBE
UNABLE TO
INTERPET SOUNDS

WERNICKE'S AREA
APHASIA/DIFFICULTY
UNDERSTANDING LANGUAGE

Source: CDC with adaptations from Stephanie Fischer

brainstem, to apathy, inability to move, disorientation, lack of restraint in behavior, social disinhibition, inability to plan, inability to feel hunger or satiety, and personality alterations as well as loss of intellectual capabilities. In a frightening number of cases the complications included death or suicide of the patient.

This treatment for otherwise-untreatable (at the time) mental illness produced some patients that were improved, some that were worse off or dead, and some that had exchanged one set of mental problems for another set of problems. This treatment came into popularity in the 1930s and persisted until the 1960s, when more psychotherapeutic drugs became available to treat more types of mental illness. The personality changes in the patients, who might become unable to restrain their behavior and might lose the very personality traits that defined them as individuals, came to inform the mythology of zombies in popular culture.

But again, the results of an identical surgery on two different people might have wildly different results, with one patient becoming unable to restrain his behavior, while another might become so apathetic she would do nothing but sit still. Therefore, lobotomy or damage to the frontal lobe of the brain is probably not a very reliable way to generate raging zombies that are specifically focused on killing others or obtaining human brains.

In fact, if you look at figure 4.3, if some sort of chemical or microbe damaged a specific lobe of the brain (for example, parts of the frontal lobe), the patient might lose the ability to speak, but retain the ability to understand language (not a common trait of zombies; they don't seem to usually be able to understand their victims). There are behavior changes resulting from frontal lobe damage that do fall into zombie patterns, such as inability to control impulses. But there are usually other changes resulting from frontal lobe damage, some that are described as lack of ambition or motivation. Those are distinctly anti-zombie patterns. And then there is the fantastic case of Phineas Gage.

Further evidence that the cerebral cortex has been damaged? Take zombies' seeming inability to feel pain. Our sense of pain stems from the somatosensory cortex of the parietal lobe (see figure 4.3). What about their apparent lack of intelligence? Frontal lobe. But they do appear to see and smell, so the occipital and temporal lobes must be intact.

There are inner sections of the cerebrum to consider as well. The limbic system controls primitive emotions and drives. The hippocampus lies at the cerebrum's base and controls both emotion and the transfer of information from short-term memory to long-term memory. Damage or decomposition of the hippocampus could have effects on cognition and language abilities. Alzheimer's disease first attacks the hippocampus resulting in memory loss and disorientation. People with damage to the hippocampus also may suffer from anterograde amnesia, losing the ability to form new long-term memories. A fictionalized example of this condition is the protagonist from the movie *Memento*.

Zombies appear to have lost their long-term memory but still retain short-term memory and also can get disoriented or distracted easily. Perhaps whatever causes zombieism damages or decomposes the hippocampus as well as select areas of the cerebral cortex.

## THE CEREBELLUM

As stated earlier the cerebral cortex is responsible for voluntary movement. But the cerebellum processes instructions to control your coordination, equilibrium, posture, and accuracy. Can you type? Play piano? Play a riff on a guitar? Thank your cerebellum. Like walking? Jogging? Skipping, dancing, prancing, strolling, sashaying? Thank your cerebellum.

Most zombies in popular culture don't have nearly the dexterity to dance the cha cha or keep a hacky sack in the air. The lurching, shambling, uncoordinated shuffling more akin to *The Walking Dead* suggest damage to the cerebellum (Kirkman, 2003-present). But the fast zombies from *28 Days Later* or *Dawn of the Dead* remake? Those zombies have a cerebellum that seems to be intact (Boyle, 2002; Snyder, 2004).

## THE BRAINSTEM

Finally we have the brainstem, which includes the midbrain, the pons, and medulla oblongata. The brainstem is in charge of your most primitive functions, regulating heart rate, breathing, and sleeping. It is also the relay station between the spinal cord and the cerebrum and cerebellum. Pain, pressure, touch, proprioception (sense of body position and movement) all go through the brainstem. The pons, for instance, relays signals from the cerebrum and cerebellum for movement.

# THE CURIOUS CASE OF PHINEAS GAGE

Phineas Gage (1823 -1860) was a foreman for the Rutland & Burlington Railroad. On September 13th, 1848 he was involved in an accident while blasting rock to make way for the rails. A tamping iron used to push charges into holes drilled into rock was blown through his lower left jaw and out of the center of the frontal bone. Miraculously he eventually made a full physical recovery save for the loss of function in his left eye.

His case led to a fundamental understanding, that personality and "mind" were one with the brain. However, how localized personality truly is, and how it can be altered, is still a confusing part of his story... a complicated amalgam of what we know about the effects of his accident (very little), and what was assumed by generations of doctors and scientists based upon their preconceived ideas about how the prefrontal cortex works.

For instance, by all accounts Gage transformed from being quite pleasant and polite before the accident to rude, coarse and generally ill tempered after. It is possible the violent animalistic nature of zombies is due in part to damage or decomposition to the prefrontal cortex.

**FIGURE 4.4.** _____ Ouch.

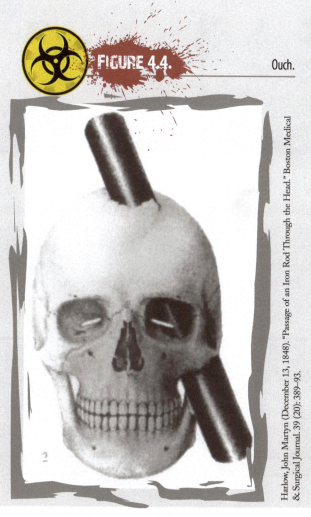

Harlow, John Martyn (December 13, 1848). "Passage of an Iron Rod Through the Head." Boston Medical & Surgical Journal. 39 (20): 389–93.

 FIGURE 4.5.

The limbic system: putting the munchies in your monsters.

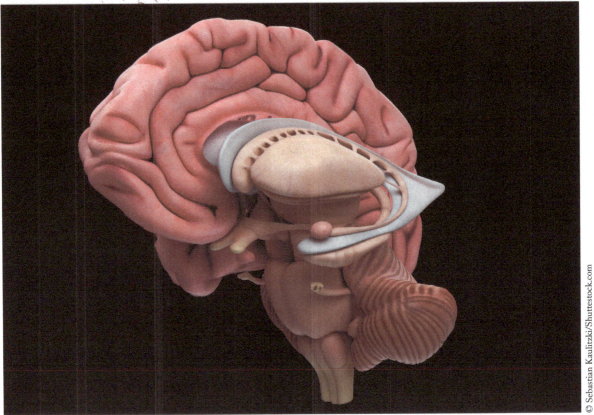

© Sebastian Kaulitzki/Shutterstock.com

As for our zombies, if any part of their brains remains fully intact, it would likely be the limbic system and brainstem. This is sometimes called the old, "reptilian" part of the brain, dedicated to survival functions like basic movement and feeding behaviors. The shambling described from damage to the cerebellum? The brainstem's pons can manage that. Zombies focused only on a food source? This part of the brain can just about handle that. And what about sight and hearing? The temporal and occipital lobes can't work properly without the midbrain. In *The Walking Dead* TV show Season 1 episode 6 it's shown that the brainstem is the only part to reawaken after death (Ferland, 2010).

**MOVIE SPOTLIGHT:** *Plan 9 from Outer Space* (Wood, 1959)
Causative agent—stimulation of pituitary and pineal glands
Written, produced, and directed by filmmaker Ed Wood
Last movie featuring horror icon Bela Lugosi (also starred in White Zombie)
"Plan 9? Ah, yes. Plan 9 deals with the resurrection of the dead. Long distance electrodes shot into the pineal and pituitary gland of the recently dead."

# WALKING CORPSE SYNDROME

Can damage to the brain really give you characteristics of a zombie? Turns out, there is a rare type of mental illness known as Cotard delusion, also known as walking corpse syndrome or walking dead man syndrome. Afflicted patients with severe cases of walking corpse syndrome actually believe they are dead or even dead and rotting. Patients often are co-diagnosed with schizophrenia. They will often refuse to eat and hallucinate, experiencing a distorted reality. Most also experience a preoccupation with self-harm and death. Others experience the symptoms only in limbs or organs, convinced these body parts are rotting, or in some cases, aren't even there.

© Artem Furman/Shutterstock.com

Walking corpse syndrome seems to be triggered by cases of brain trauma, tumors, infections, or strokes. Epileptic seizures also seem to be associated with the disease. It can be treated with antipsychotics, antidepressants, or even electroconvulsive therapy (ECT).

**FIGURE 4.6.** Practice makes perfect.

# NEURONS AND NEUROTRANSMITTERS

Next, let's look at the microscopic level of brains. Brains are composed of specialized cells called neurons. They are truly unique among cells. Neurons can conduct signals both electrically (charge) and chemically (neurotransmitters). Let's take a closer look at a connection between a transmitting neuron and a receiving neuron (figure 4.7).

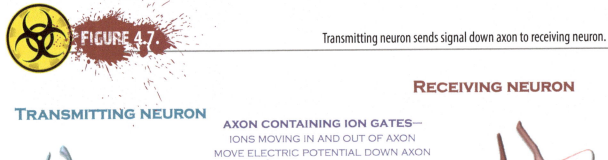

**FIGURE 4.7.**

Transmitting neuron sends signal down axon to receiving neuron.

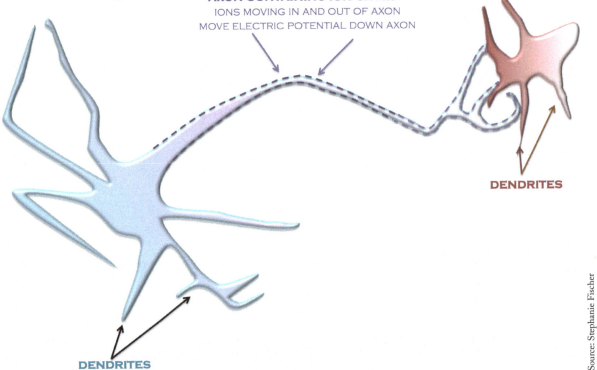

**RECEIVING NEURON**

**TRANSMITTING NEURON**

**AXON CONTAINING ION GATES—**
IONS MOVING IN AND OUT OF AXON
MOVE ELECTRIC POTENTIAL DOWN AXON

**DENDRITES**

**DENDRITES**

Source: Stephanie Fischer

The neuron contains dendrites which receive chemical signals either from a receptor or another neuron. They convert this signal to an electrical signal and send it to the cell body, or soma. The cell body receives a summation of all the signals from dendrites and determines if the signals are strong enough to pass on to the next cell.

The cell body will then send the signal down the single axon. The axon contains ion gates, which open when an electrical potential reaches them, allowing more ions (charged atoms or molecules) in, and continuing the wave of ions. It's almost like the "wave" you see at a sports event. Ions move in and then back out, creating a wave of positive charge called an electrical potential that moves down the axon. The axon is wrapped in a myelin sheath made up of Schwann cells that insulates the axon and greatly speeds up the signal. (This is pretty oversimplified, but enough to know for a zombie class.)

**FIGURE 4.8.**

Why, hello to you, too.

©Joseph Sohm/Shutterstock.com

Let's briefly think about another undead creature from our genre: Frankenstein's monster (Shelley, 1818). At the time *Frankenstein* was written, scientists were just beginning to play with and understand the role of this electrical potential in brains and muscles. They were able to take dead animals and apply an electrical current, and get the animal's muscles to twitch. Some of these people were more entertainers than scientists or doctors, and claimed they brought the animals back to life. In reality, they were applying an electrical charge to a wiring system that still existed after death (but had no energy to function on its own). It was pretty frightening for people at the time, and gave rise to the classic Frankenstein story and many to follow, wondering what would happen if you really could bring a corpse back to life with a bolt of lightning or other electrical impulse.

Of course, the problem with this hypothesis is, how do you sustain the electrical potential? We talked about ATP and energy carriers in the brain and body in the previous section. Think of it like this… your cell phone has been damaged by a virus, which corrupted its software programs and caused it to "brick," or die. Will plugging it in help? Powering the *hardware* isn't the same as *repairing the software*. The phone can have power, but if the instructions that run the software aren't functional, the phone won't magically regain function just because you plug it in. You can run electricity through a dead body and get some muscle twitches (for a while, till the proteins decay). However that's not the same as returning the body to life, making it self-sustaining, and restoring brain function. Ok, back to zombies.

So we know the basics of how the axon works as an electrical conductor. Now we travel to the end of the axon at the axon terminals. There is a space between the transmitting neuron and the receiving neuron, called a synaptic cleft (figure 4.9). To jump across the cleft, the electrical potential is converted to a chemical signal, called a neurotransmitter.

Neurotransmitters are also how neurons from the nervous system tell muscle cells when to contract. They control our respiratory rate and our heart rate. They control our sleep and wake cycles. They control how we feel pain.

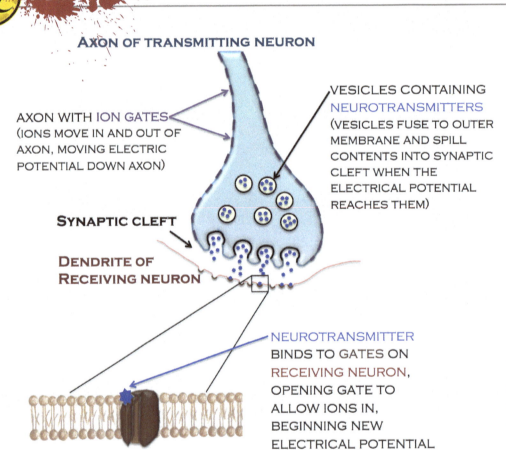

**FIGURE 4.9.** Neurotransmission across a synapse.

**AXON OF TRANSMITTING NEURON**

AXON WITH ION GATES (IONS MOVE IN AND OUT OF AXON, MOVING ELECTRIC POTENTIAL DOWN AXON)

VESICLES CONTAINING NEUROTRANSMITTERS (VESICLES FUSE TO OUTER MEMBRANE AND SPILL CONTENTS INTO SYNAPTIC CLEFT WHEN THE ELECTRICAL POTENTIAL REACHES THEM)

**SYNAPTIC CLEFT**

**DENDRITE OF RECEIVING NEURON**

NEUROTRANSMITTER BINDS TO GATES ON RECEIVING NEURON, OPENING GATE TO ALLOW IONS IN, BEGINNING NEW ELECTRICAL POTENTIAL

Source: Stephanie Fischer

## SEROTONIN

There are many neurotransmitters as well as different receptors to read them. The receptors can be excitatory (beginning a new electrical potential), inhibitory (making it less likely for a new electrical potential to start), or modifying (any range between the two). In addition, a single neurotransmitter, like serotonin, can be excitatory or inhibitory depending on the receptor and the other inputs. The versatility of serotonin's interaction with the different receptors means the neurotransmitter can perform a wide range of functions. From the brain stem, concentrations of serotonin-based neurons project to the rest of the brain, to do many things: inhibiting pain sensation, controlling sleep cycles, controlling alertness, and modifying appetite, among others. In addition, it is believed that the primary function of serotonin is to affect your mood, increasing your feelings of happiness and well-being.

There is some evidence that people with chronic depression have reduced levels of serotonin in synapses in their brains. In some people, this seems to be because of a genetic alteration in the serotonin reuptake transporter (sometimes spelled "re-uptake"; figure 4.10). A neurotransmitter reuptake transporter pulls the original neurotransmitter back out of the synaptic cleft and into the transmitting neuron to be reused. It basically recycles the neurotransmitter to be used again.

**FIGURE 4.10** Reuptake transporters recycling serotonin back into transmitting neuron.

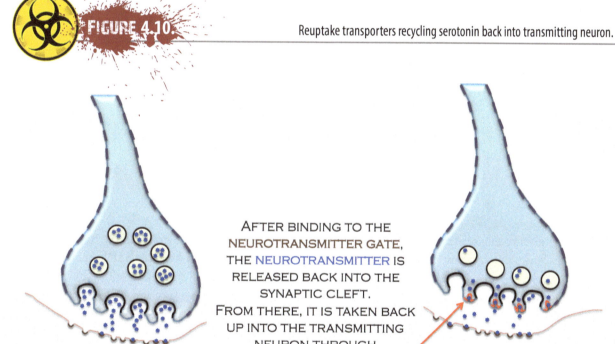

AFTER BINDING TO THE NEUROTRANSMITTER GATE, THE NEUROTRANSMITTER IS RELEASED BACK INTO THE SYNAPTIC CLEFT. FROM THERE, IT IS TAKEN BACK UP INTO THE TRANSMITTING NEURON THROUGH REUPTAKE TRANSPORTERS

Source: Stephanie Fischer

If the reuptake transporter is too active, a person will not get a strong enough serotonin signal because the serotonin will not remain in the synapse long enough to signal to the receiving neuron. It is as though someone is grabbing your coffee cup out of your hand before you finish the coffee and chucking the cup into the recycling bin. You're going to stay thirsty (and sleepy).

Inversely, if the transporter is not active enough, the person will get too strong a serotonin signal, because the serotonin remains in the synapse too long, flooding the synapse, and constantly activating the receiving neuron (figure 4.11). It is as though you're surrounded by every coffee cup from every coffee you've had this month… and there are no recycling bins!

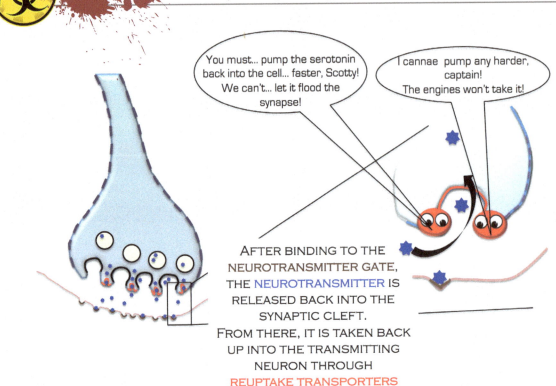

Source: Stephanie Fischer

There is some scientific evidence that a person with too active a transporter is more likely to have clinical depression. Some patients respond well to a selective serotonin reuptake inhibitor (SSRI) that slows the transporter down, so more serotonin is left in the synapse. Whether a patient responds well to this treatment depends on his or her genes, because small differences in the transporter structure can change how it works and how well it responds to antidepressant drugs. There is a lot of research still going on about how diverse the interactions of neurotransmitters are, and how individual people's responses are. The complexities are beyond the scope of this text, but sometimes oversimplifications can help explain things in general! [Lazarevic reference]. We will be talking about genetics and protein function in the next section.

Other main neurotransmitters modify and balance these effects, so that really the entire brain is in a balancing act. A Venn diagram (figure 4.12) can explain this (again, it is oversimplifying a bit, but gives the right idea).

**BOOK SPOTLIGHT:** *Magic Island* (Seabrook, 1929)
Offers firsthand accounts of Haitian voodoo.
Credited with introducing the concepts of zombies to western culture; Inspired the first zombie movie, *White Zombie* (Halperin, 1932)
Follow's the author's travelogue to Haiti; Narrative is filled with racial prejudices toward Haitians, especially the illustrations by Alexander King.

**FIGURE 4.12**

Interaction of primary neurotransmitters.

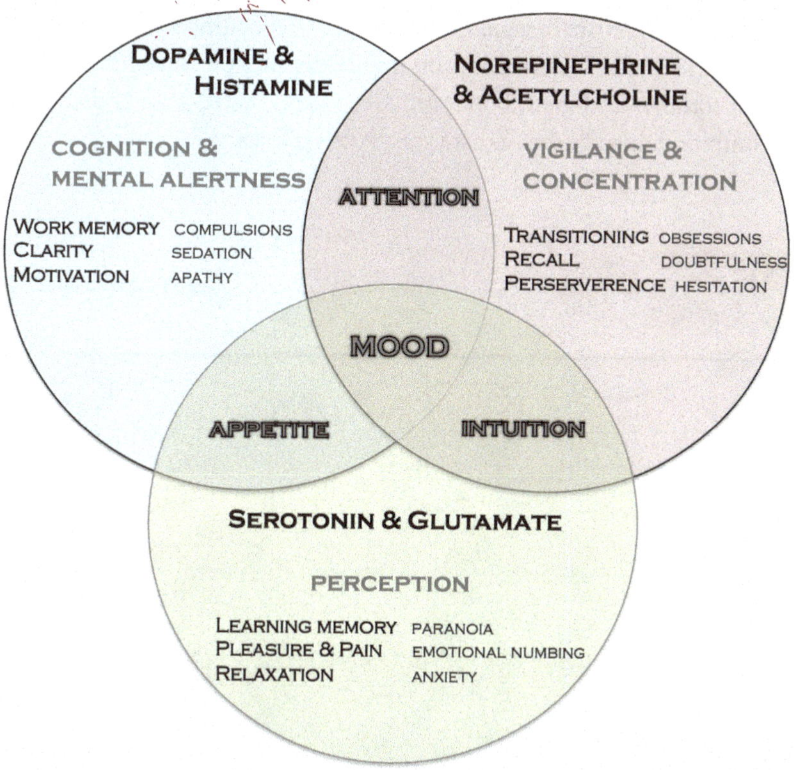

**DOPAMINE & HISTAMINE**

COGNITION & MENTAL ALERTNESS

WORK MEMORY    COMPULSIONS
CLARITY        SEDATION
MOTIVATION     APATHY

**NOREPINEPHRINE & ACETYLCHOLINE**

VIGILANCE & CONCENTRATION

TRANSITIONING  OBSESSIONS
RECALL         DOUBTFULNESS
PERSERVERENCE  HESITATION

ATTENTION

MOOD

APPETITE

INTUITION

**SEROTONIN & GLUTAMATE**

PERCEPTION

LEARNING MEMORY  PARANOIA
PLEASURE & PAIN  EMOTIONAL NUMBING
RELAXATION       ANXIETY

Source: Stephanie Fischer

One oversimplification we often see in popular literature is that serotonin makes you feel happy. Well, it maybe has more to do with contentment. It does seem that people with slightly higher (but not too high) levels of serotonin released into their synapses at the appropriate time, do report higher feelings of contentment. But is it really that simple?

In the apocalyptic 2002 hit movie *28 Days Later*, the causative agent (what started the illness) was a genetically modified "Rage" virus. In the movie the virus was engineered to inhibit neurotransmitters that cause rage and violence, but when tested it had the reverse effect and instead amplified rage and violence in the infected.

MOVIE SPOTLIGHT: *28 Days Later...* (Boyle, 2002)
Causative agent—genetically modified "Rage" virus
Commercial and critical success credited with reviving the zombie genre
"Zombies" are infected humans, not reanimated corpses
"It started as rioting. But right from the beginning you knew this was different. Because it was happening in small villages, market towns. And then it wasn't on the TV any more. It was in the street outside. It was coming in through your windows."

Viruses and microorganisms will be covered in detail in the next chapter. But for now let's just focus on the idea that you can "isolate" the neurotransmitters that cause rage and violence. So, the premise is that this "rage virus" somehow destroys serotonin release, making the affected patients "filled with rage." The problem with that is, it just isn't that simple. A patient with no serotonin (just looking at emotional outcomes) would not be consumed with rage, but more likely consumed with inertia and not want to do anything. They would also feel pain very acutely, be highly distractible, and have no appetite. So blocking serotonin doesn't give us zombies that are full of rage. More like… well, dead bodies that sit around and rot. Which… is what unzombified dead bodies do anyway. Not very scary.

FIGURE 4.13.                                        Zombie with no serotonin.

© Emily Adele

## DOPAMINE

What about another neurotransmitter? High levels of dopamine contribute to feelings of ecstasy and reward, clear thinking, and hunger. If we block dopamine, could we create zombies? Well, they wouldn't be so much full of rage as full of disappointment, and they might be very confused. They would be pretty unmotivated to get out and do much of anything, although if eating others' brains gave them a dopamine rush, they might be willing to get out there and find some brains to restore their reduced dopamine levels. Of course, eating anything digests the proteins before making it out of the digestive system, so that wouldn't work to restore very high levels of dopamine in zombie brains. They'd have to process the brains of their victims and then somehow get that brain extract into their bloodstreams or central nervous systems. Zombies snorting their victim's brains? Weird. And how would zombies separate out dopamine from their victims other neurotransmitters? They'd be better off looking for pure dopamine-type drugs, as we use for Parkinson's patients. So, inhibiting the effects of dopamine wouldn't create textbook zombies either. It'd more likely create zombies that would raid pharmacies for dopamine drugs.

## Acetylcholine

As for the third group in our picture, if you block acetylcholine you get paralysis, because that is the neurotransmitter your nerves use to communicate with your muscles. Zombies paralyzed like a marble statue aren't very scary. So that's not going to work, either.

Actually, all of these scenarios have been a bit oversimplified. Any one of our emotions isn't caused by just one neurotransmitter. It is really more like a symphony of neurotransmitters being released and taken up, all over the brain. The specific area of the brain that is receiving the signal is just as important, since the different areas can modify how those signals are manifested. Remember the functions of the different sections of cerebral cortex.

## NEUROTOXINS

Let's get back to where we started this chapter, Haiti and the neurotoxin-induced zombies. Neurotoxins are chemicals that specifically affect neurons. Some neurotoxins affect all neurons equally, and some preferentially damage dopamine neurons, or serotonin neurons, for example.

So could neurotoxins create zombies? Well, again, we run into the problem about whether you can knock out one type of neuron and get the mindless zombie result. Most neurotoxins would look rather a lot like eliminating a specific neurotransmitter, like serotonin—and we already saw how that doesn't produce our mindless and homicidal zombies.

Could there specifically be a neurotoxin that just damaged the frontal lobe, for example? Maybe. But then… the neurotoxin would exist out there somewhere, and wouldn't be infectious. But it could be some kind of environmental exposure, perhaps?

Now, this could happen. Many neurotoxins (like lead, mercury, and animal poisons) affect those ion channels our neurons use to generate electrical signals. Other neurotoxins (like botulism toxin and tetanus toxin) interfere with the release or reuptake

**FIGURE 4.14.** Zombie with no dopamine.

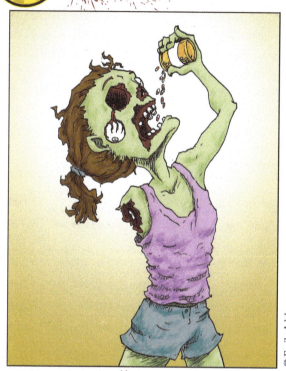

**FIGURE 4.15.** Zombie with no acetylcholine.

© Emily Adele

© Emily Adele

of neurotransmitters. But usually, the effects are so widespread that the toxins don't create a mindless person… just a dead one. Is it possible that there is a neurotoxin out there (or that one could be designed) to create mindless people? Well, yeah. Our anesthetics normally shut down our conscious mind so that we don't feel pain, but they don't kill us (hopefully). We have types of drugs that remove people's inhibitions, cause them to sleep-walk, or make them act strangely. We call those mind-altering drugs. So that's probably our best bet at getting zombies… but then, usually those drugs wear off, and patients are normal again.

"Tetanus Following Gunshot Wounds," 1809. By the surgeon and artist Sir Charles Bell, the painting depicts a soldier suffering from tetanus.

But neurotoxins affecting regions of the brain *do* get us as close as we've gotten so far to "mindless" humans….

| Neurotoxin | Source | Neural effect | Signs and symptoms |
|---|---|---|---|
| Anatoxin-*a* | cyanobacteria (blue green algae) | ACh receptor blocker | twitching, convulsions, death from respiratory paralysis |
| Botulinum toxin | *Clostridium botulinum* bacteria | ACh inhibitor | flaccid paralysis |
| Saxitoxin | cyanobacteria (blue green algae) | sodium channel blocker in axons | flaccid paralysis |
| Tetanus toxin | *Clostridium tetani* bacteria | ACh reuptake inhibitor | muscle spasms that may result in bone fractures |
| Tetrodotoxin | order Tetradontiformes (including pufferfish) | sodium channel blocker in axons | prevents nervous system from sending commands, flexing muscles |

## THE QUESTION STILL REMAINS, HOWEVER…

Could damage to the brain cause zombies? From the limited evidence we have (based on when lobotomies were performed, for example)… the answer to that is a qualified yes. Sometimes it might; sometimes, it just might kill the person. Sometimes, it wouldn't affect them much at all. It could make them "mindless," like zombies, but also would leave them unmotivated to do much like move, eat, or drink. It could make them more loving or make them apathetic. Sometimes it might affect their ability to communicate, and sometimes it wouldn't… so it wouldn't be a reliable and repeatable way to make zombies.

So far, we've explored how neurons work, and how manipulating one or two neurotransmitters could definitely interfere with brain function, but would not give you classic zombies. At least, not zombies that can still move, but would be devoid of intellect and personality, because so much of how neurotransmitters control brain and body function are integrated in very complex ways.

In addition, we've looked at how some brain functions are localized into specific areas, and how damage to one area (like the frontal lobe) can have dramatic effects on personality, but often also destroys motivation as well as ability to move.

So, neither messing with neurotransmitters nor damaging parts of the brain will give you our zombies. A localized brain modification seems to be an unreliable way to create our zombies. Blocking or replacing a single neurotransmitter won't create our zombies.

On top of that, even if there *was* a chemical that could affect your brain and make you a zombie, you wouldn't be contagious. So that would preclude any zombie apocalypse. So perhaps we should look elsewhere.

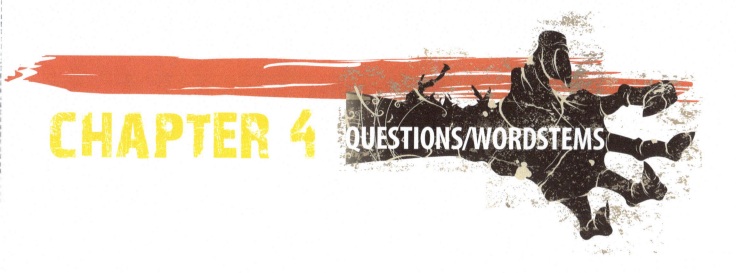

# CHAPTER 4 — QUESTIONS/WORDSTEMS

1. The long part of a neuron that reaches out towards a receiving neuron is called the
   a. Axon.
   b. Synapse.
   c. Dendrite.
   d. Ion.
   e. Gate.

2. The space between two neurons that must be crossed by a neurotransmitter is called a
   a. Synapse.
   b. Axon.
   c. Gate.
   d. Ion.
   e. Dendrite.

3. The signal that travels across the space between two neurons, comprised of packets of neurotransmitters released across the space (like balloons), is the _____ portion of the neuron signal.
   a. Chemical
   b. Electrical
   c. Boring
   d. Excitatory

4. In the scientific literature, did patients receiving lobotomies consistently lose ability to communicate? Explain your answer.

5. Explain why a patient missing all serotonin does not become "full of rage."

| WORDSTEMS | |
|---|---|
| -campus | sea creature |
| ethno- | human related, nation |
| hippo- | horse |
| -mancy | divination or magic |
| necro- | dead |
| occipit- | back of head |
| neuro- | nerve |
| syn- | together, like |
| -tomy | cut or remove |
| toxi- | poison |

# CHAPTER 4

**WORSHEET**

## CAUSATIVE AGENTS IN MOVIES (GROUP PROJECT)

### PART A

Get together with your group and discuss the zombie movies you've seen. List the causative agents of infection (if known) as chemical, radiation, microorganism, or religious/supernatural and other malarkey. Is it a hybrid cause? Several come to mind for me.

### PART B

Pick one of these movies you would like to advocate as the zombie movie we will watch (and pick apart) in class. Don't pick one with a religious/supernatural causative agent. As a group or with a single spokesman, give a 2–4 minute pitch for the movie you want to watch later in the semester. I would like at least four MS PowerPoint slides for the presentation. I will assist each group that isn't familiar with the software.

The winning movie will be chosen by me, but the class vote will be a major factor (I have to consider content, scientific feasibility, and whether it has been screened previous semesters). The movie will also be summarized by the class in a later worksheet, then retired. So no matter how many times this gets taught, this movie is *your* zombie movie.

_____

_____

_____

_____

_____

_____

_____

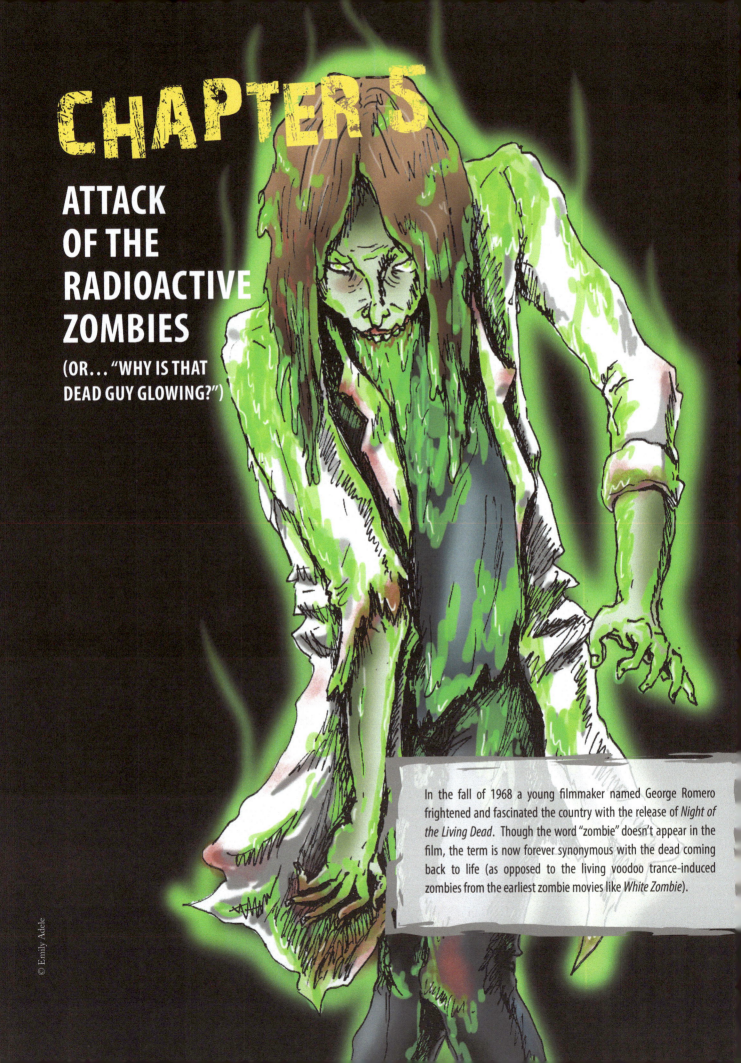

# CHAPTER 5

## ATTACK OF THE RADIOACTIVE ZOMBIES

**(OR... "WHY IS THAT DEAD GUY GLOWING?")**

In the fall of 1968 a young filmmaker named George Romero frightened and fascinated the country with the release of *Night of the Living Dead*. Though the word "zombie" doesn't appear in the film, the term is now forever synonymous with the dead coming back to life (as opposed to the living voodoo trance-induced zombies from the earliest zombie movies like *White Zombie*).

Images: Public Domain

 **FIGURE 5.1.**

Images from *Night of the Living Dead* (Romero, 1968).

**MOVIE SPOTLIGHT:** *Night of the Living Dead* (Romero, 1968)
Causative agent—unknown, suspected as radioactive contamination from a space probe
Heavily criticized at the time of release for explicit gore
Never used the word "zombie." Undead were referred to as "ghouls"
"It has been established that persons who have recently died have been returning to life and committing acts of murder."

Even though the cause of the zombie outbreak is never revealed in *Night of the Living Dead*, it probably isn't a coincidence that the resurgence of zombies as a horror genre really started at the same time the world was coming to terms with high-radiation weapons and energy sources (1968, the year of the release of *Night of the Living Dead*, was also the year the Non Proliferation Treaty was signed, wherein several countries with nuclear capability agreed to build no more and to begin to disarm, and countries without nuclear capability agreed to not build any). The world had begun to be horrified at the destruction such radiation could cause. Several movies since have also pointed to radiation as the culprit.

> "Because of the obvious threat to untold numbers of citizens due to the crisis that is even now developing, this radio station will remain on the air day and night. This station and hundreds of other radio and TV stations throughout this part of the country are pooling their resources through an emergency network hook-up to keep you informed of all developments. At this hour, we repeat, these are the facts as we know them. There is an epidemic of mass murder being committed by a virtual army of unidentified assassins. The murders are taking place in villages and cities, in rural homes and suburbs with no apparent pattern nor reason for the slayings. It seems to be a sudden general explosion of mass homicide. We have some descriptions of the assassins. Eyewitnesses say they are ordinary-looking people. Some say they appear to be in a kind of trance. Others describe them as being misshapen monsters." (Romero, 1968)

But can radiation really cause a zombie outbreak? Well, exposure to radiation can damage or alter DNA. Changes in DNA are called mutations, which can profoundly affect how an organism functions. Let's see how that happens.

# GENETIC INFORMATION AND HOW IT FLOWS FROM DNA TO PROTEIN

All cells contain DNA, a macromolecule with the ability to store and transmit information. DNA is made up of individual monomers called nucleotides, each with an unchanging "backbone" of phosphates and sugars, and then has four different nitrogenous bases towards the middle (Adenine, Cytosine, Guanine, and Thymine, or A, C, G, and T). It is the specific sequence of the bases that holds the genetic information (figure 5.2).

DNA is actually a double-stranded helix; two parallel strands of nucleotides are connected at their nitrogenous bases with the backbones facing outwards. These two strands twist around each other forming a helix. Picture a rope ladder twisted over and over. Most importantly, the nitrogenous bases only fit together in a specific order. Thymine will only bind to Adenine, while Guanine will only bind with Cytosine. So if one strand reads ATCCGTACG, for instance, the corresponding strand must read TAGGCATGC. This pairing is actually the keystone to holding genetic information and passing it on to the next generation.

Some people wonder how information can be stored in sequences of just four bases… but your cell phone and your computer function in binary—that is, all the information in your cell phone is stored as sequences of 1s and 0s. (Really, positive or negative charges, but this class is all about keeping things simple.)

 **FIGURE 5.2.**

Nucleotides and DNA.

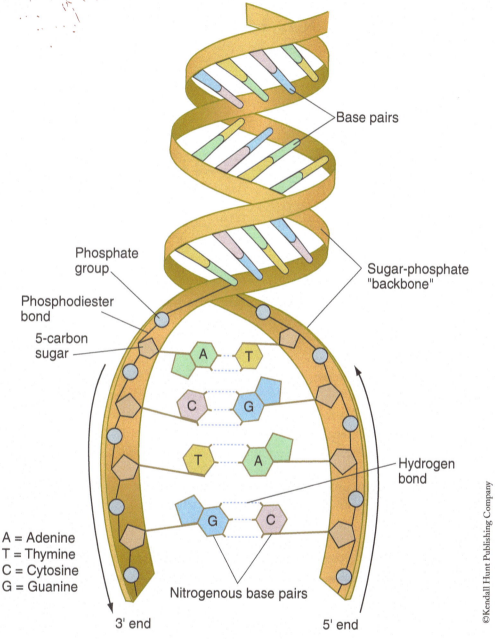

Base pairs

Phosphate group

Phosphodiester bond

5-carbon sugar

Sugar-phosphate "backbone"

Hydrogen bond

A = Adenine
T = Thymine
C = Cytosine
G = Guanine

Nitrogenous base pairs

3' end

5' end

©Kendall Hunt Publishing Company

For other people, it makes more sense to think of music, where a limited number of notes can, in different sequences, compose such different information as Beethoven's *Moonlight Sonata* (Figure 5.3) or Beyoncé's *Single Ladies*.

FIGURE 5.3.

## Sonata No. 14 "Moonlight"
### 1st Movement

L. VAN BEETHOVEN
Op. 27, No. 2

ADAGIO SOSTENUTO

*sempre* **pp** *e senza sordino*

**pp**

© Christopher Green

It's true that DNA contains the genetic information for every physical and physiological trait you possess. It essentially contains all that makes you… you. Now that may sound grandiose, but think of your DNA as a giant, multivolume cookbook. It contains the recipes for making thousands of proteins. And if it is making proteins, it is making enzymes. Don't forget, enzymes are the tools that build up or break down every part of your cell. So it's more accurate to say your DNA contains the information to build the *tools* that make you… you.

A sequence of DNA that contains the information for a single protein is called a gene. For example, the insulin gene is 2764 bases long (with 1431 of those bases actually holding the information for the protein; figure 5.4).

Once that information is transcribed (into another form of nucleic acid, mRNA), it is then translated (used as instructions to make) into a protein. The insulin protein helps your cells import glucose (a sugar) as fuel to make ATP.

Proteins are the molecules that really do all the WORK in your cells… they are the enzymes, and the cytoskeleton, the transport systems, and the communication molecules. Proteins pretty much do everything; and genes are the instructions to make those proteins.

One of the best metaphors we have been able to come up with to explain DNA, genes, mRNA, and proteins is using Legos.

**FIGURE 5.4.**

Genetic code for insulin.

```
gggcctcagctggggctgctgtcctaaggcagggtgggaactaggcagccagcagggagg        60
ggacccctccctcactcccactctcccacccccaccaccttggcccatccatggcggcat       120
cttgggccatccgggactggggacaggggtcctggggacaggggtgtggggacaggggtc       180
ctggggacaggggtctggggacaggggtcctggggacaggggtgtggggacaggggtgtg       240
gggacaggggtgtggggacaggggtcctggggacaggggtctggggacaggggtctgagg       300
acaggggtgtggggacaggggtgtggggacaggggtgtggggacaggggtgtggggacag       360
gggtctggggacaggggtccggggggacaggggtgtggggacaggggtgtggggacagggg       420
tgtggggacaggggtctggggacaggggtgtggggacaggggtcctggggacaggggtgt       480
ggggatagggtgtggggacaggggtgtggggacaggggtgtggggacaggggtctgggg       540
acagcagcgcaaagagccccgccctgcagcctccagctctcctggtctaatgtggaaagt       600
ggcccaggtgagggctttgctctcctggagacatttgcccccagctgtgagcagggacag       660
gtctggccaccgggcccctggttaagactctaatgacccgctggtcctgaggaagaggtg       720
ctgacgaccaaggagatcttcccacagacccagcaccagggaaatggtccggaaattgca       780
gcctcagcccccagccatctgccgacccccccaccccaggccctaatgggccaggcggca       840
ggggttgagaggtaggggagatgggctctgagactataaagccagcggggggcccagcagc       900
cctcagccctccaggacaggctgcatcagaagaggccatcaagcaggtctgttccaaggg       960
cctttgcgtcaggtgggctcaggattccaggtggctggaccccaggccccagctctgca      1020
gcagggaggacgtggctgggctcgtgaagcatgtggggggtgagcccaggggccccaaggc      1080
agggcacctggccttcagcctgcctcagccctgcctgtctcccagatcactgtccttctg      1140
ccATGGCCCTGTGGATGCGCCTCCTGCCCCTGCTGGCGCTGCTGGCCCTCTGGGGACCTG      1200
ACCCAGCCGCAGCCTTTGTGAACCAACACCTGTGCGGCTCACACCTGGTGGAAGCTCTCT      1260
ACCTAGTGTGCGGGGAACGAGGCTTCTTCTACACACCCAAGACCCGCCGGGAGGCAGAGG      1320
ACCTGCAGGgtgagccaactgcccattgctgcccctggccgcccccagccacccctgct      1380
cctggcgctcccacccagcatgggcagaaggggggcaggaggctgccacccagcagggggt      1440
caggtgcactttttaaaaagaagttctcttggtcacgtcctaaaagtgaccagctccct      1500
gtggcccagtcagaatctcagcctgaggacggtgttggcttcggcagccccgagatacat      1560
cagagggtgggcacgctcctccctccactcgcccctcaaacaaatgccccgcagcccatt      1620
tctccaccctcatttgatgaccgcagattcaagtgtttttgttaagtaaagtcctgggtga      1680
cctggggtcacaggggtgccccacgctgcctgcctctgggcgaacaccccatcacgcccgg      1740
aggagggcgtggctgcctgcctgagtgggccagacccctgtcgccaggcctcacggcagc      1800
tccatagtcaggagatggggaagatgctggggacaggccctggggagaagtactgggatc      1860
acctgttcaggctcccactgtgacgctgcccccggggcggggaaggaggtgggacatgtg      1920
ggcgttggggcctgtaggtccacacccagtgtgggtgaccctccctctaacctgggtcca      1980
gcccggctggagatgggtgggagtgcgacctagggctggcgggcaggcgggcactgtgtc      2040
tccctgactgtgtcctcctgtgtccctctgcctcgccgctgttccggaacctgctctgcg      2100
cggcacgtcctggcagTGGGGCAGGTGGAGCTGGGCGGGGGCCCTGGTGCAGGCAGCCTG      2160
CAGCCCTTGGCCCTGGAGGGGTCCCTGCAGAAGCGTGGCATTGTGGAACAATGCTGTACC      2220
AGCATCTGCTCCCTCTACCAGCTGGAGAACTACTGCAACtagacgcagcccgcaggcagc      2280
cccacacccgccgcctcctgcaccgagagagatggaataaagcccttgaaccagccctgc      2340
tgtgccgtctgtgtgtcttggggggccctgggccaagccccacttcccggcactgttgtga      2400
gcccctccagctctctccacgctctctgggtgcccacaggtgccaacgccggccaggcc      2460
cagcatgcagtggctctccccaaagcggccatgcctgtcggctgcctgctgcccccaccc      2520
tgtggctcagggtccagtatgggagctgcggggggtctctgaggggccaggggtggtgggg      2580
ccactgagaagtgacttcttgttcagtagctctggactcttggagtccccagagaccttg      2640
ttcaggaaagggaatgagaacattccagcaattttccccccacctagccctcccaggttc      2700
tatttttagagttatttctgatggagtccctgtggagggaggaggctgggctgagggagg      2760
gggt                                                              2820
```

FIGURE 5.5.

A Corellian freighter of some renown.

©Marek Szandurski/Shutterstock.com

Somewhere is a collection of every instruction for every set Lego has ever built, right? That original, all-inclusive hard copy library of instructions on how to build every Lego ship, set, craft, or vehicle is all made of pictures and words (our DNA), and each individual set of instructions is a gene. Genes are made of pieces of DNA that contain the instructions for one protein.

But when you get a Lego set, they don't send you the original blueprints to all Lego sets ever made.[1] They give you a *copy* of the instructions *for your set only*. That's the mRNA. An mRNA is a working copy of a gene.

Just as an example, let's say we are interested in the gene for making a Lego Millennium Falcon. What they ship you in the box is a working copy of those instructions, which can be lost or destroyed when you are done with it, and they will still have the original copy, safe and sound. Your set of instructions is the mRNA.

But, a set of instructions is not the same thing as the Lego Millennium Falcon. Your cells have the instructions for making proteins (DNA and RNA), but that's not the same thing as the proteins themselves. How do you end up with a super-cool Millennium Falcon you can play with? How do your cells build the proteins they need?

Something has to USE the instructions, and the raw materials (the Legos) to actually build the Millennium Falcon. In our cells, ribosomes are able to "read" the instructions in mRNA and use those instructions to build proteins. In our Lego metaphor, you are the ribosome. You are going to read the instructions and follow them to build a Lego Millennium Falcon to play with.

---

1. Yes this actually exists. Lego has an archive vault with original blueprints and models of every set ever produced!

# TRANSCRIPTION (DNA TO RNA)

So where do we start? We have a bunch of information… a sequence of nitrogenous bases that we can abbreviate as As and Ts, Cs and Gs. Well, just like those musical notes, or the letters and pictures in Lego instructions, those letters *mean* something.

A process called transcription copies those pieces of information from the big books (DNA) to smaller index cards (mRNA). A single gene (instructions for one protein) is extracted from the whole genome, and transcribed (recopied) into mRNA (figure 5.6).

So how do we know where to start reading? Well the beginning of each gene is called the promoter and is easily identified by what is called the TATA box. The TATA box refers to the sequence on the template strand (the strand of DNA you read) of the DNA starting with TATAAAA. This signifies both the beginning of the promoter and the template strand (as opposed to the non-template strand).

After identifying the TATA box we can start transcribing. We can build the RNA by reading the template strand. RNA contains Adenine, Guanine, and Cytosine (just like DNA), but has the nitrogenous base Uracil in place of Thymine. G still binds with C, but now A binds with U. So if our template strand reads ATCCGTACG, for instance the corresponding mRNA strand must read UAGGCAUGC.

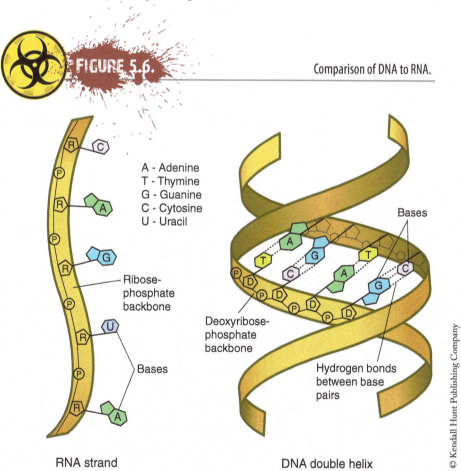

**FIGURE 5.6.** Comparison of DNA to RNA.

A - Adenine
T - Thymine
G - Guanine
C - Cytosine
U - Uracil

Ribose-phosphate backbone

Bases

RNA strand

Bases

Deoxyribose-phosphate backbone

Hydrogen bonds between base pairs

DNA double helix

## TRANSLATION (RNA TO PROTEIN)

Now the messenger RNA contains the same information the DNA did, but it is all nice and condensed (like getting that nicely published instruction set in your Lego package, or you taking notes from a long textbook and putting that one concept concisely on one index card). Remember that RNA doesn't use the same chemical thymine (abbreviated as T) as DNA. Instead, it uses uracil, or U. It's more stable that way.

The mRNA leaves the nucleus and goes into the cytosol where it is found by a ribosome. The ribosome binds to the mRNA and begins to "read" it, by matching up the bases in the messenger RNA with an amino acid. An amino acid is a chemical group that is the building block of proteins. There are 20 different amino acids, each with different characteristics (like whether it has a positive charge, a negative charge, "likes" water, or "hates" water).

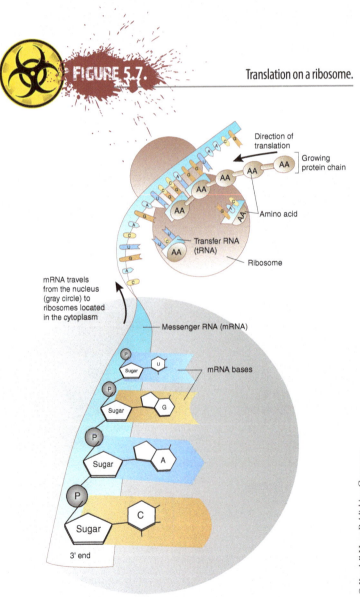

FIGURE 5.7.    Translation on a ribosome.

Direction of translation

Growing protein chain

Amino acid

Transfer RNA (tRNA)

Ribosome

mRNA travels from the nucleus (gray circle) to ribosomes located in the cytoplasm

Messenger RNA (mRNA)

mRNA bases

3' end

Ribosomes read mRNA nitrogenous bases in groups of three. If the nitrogenous are the "letters" of our genetic language then the "words" are the three base codons. Refer to the genetic code on page 89. This is the Rosetta Stone of genetic information and is a blueprint of how ribosomes work. When the ribosomes read the codons, they place the matching amino acids. At the end of this chapter, we will go over this process step by step from DNA to mRNA to finished protein.

FIGURE 5.8. Damn right, I shot first.

© Ego

The process will continue until each codon in the messenger RNA has been matched to its set amino acid, and the ribosome has linked together each amino acid in a chain in the proper order (figure 5.7).

What you end up with is a chain of amino acids.

So to summarize the Lego analogy, the gene for the Millennium Falcon contains the master list of instructions at the Lego headquarters on how to build the Corellian freighter (the exact sequence of amino acids). The mRNA for the Millennium Falcon contains a copy of the same instructions that come in the box. You become the ribosome when you build the Falcon, and the Lego bricks are the amino acids. Once you as the ribosome finish the instructions you have the completed Millennium Falcon (final protein). Now go make the Kessel run… you've got 12 parsecs.[2]

## PROTEIN SHAPE AND FUNCTION

Proteins do almost all the important work in a cell. They build things, they break things down. They are the structure of cells (like scaffolding or building frameworks), they are the train tracks for transportation, and they are the engines that move things along those tracks.

We just learned how every protein is coded for by a gene (a piece of DNA). The instructions for how to build a protein (the sequence of amino acids) is embedded in the sequences of DNA bases (the As, Ts, Cs, and Gs).

A protein is made of amino acids, in the same way that figure 5.9 is made of Legos.

2. The authors are aware of the current debate regarding the use of parsecs as a time measurement rather than distance in this context. The authors also choose to believe that Corellian smugglers understand the colloquial use of terms in various eras, in the same way we currently use "cool" to mean other things than temperature, and therefore please do not contact the authors with corrections or justifications; we get it. Really.

FIGURE 5.9.

Lego Engineer Xander Green with a full set of instructions

Source: Christopher Green

But the order you put the Legos together in is pretty important, or you get something that may be incomplete, as in figure 5.10.

FIGURE 5.10.

Lego Engineer Xander Green with the last four pages of the instructions ripped out

Source: Christopher Green

Now, some changes in the instructions might change up what the end product (protein) looks like, but they might be benign (not harmful)… or even an improvement on the shape and function of the protein, as in figure 5.11.

**FIGURE 5.11.** Lego Engineer Xander Green with random instructions taped inside over the correct instructions

Source: Christopher Green

But some changes might completely change the shape of the protein, so that it cannot function at all (figure 5.12).

**FIGURE 5.12.** Lego Engineer Xander Green with the family dog chewing the instructions to pieces

Source: Christopher Green

If that protein performs a critical function in the organism, then its lack of function can cause a disease state.

## SICKLE-CELL ANEMIA: AN EXAMPLE OF DNA MUTATION CAUSING A DISEASE

Let's look at what happens when a change in instructions for a protein has dangerous consequences.

Hemoglobin is a carrier protein with four bound iron ions ($Fe^{2+}$); these are responsible for holding onto oxygen and carrying it to your cell so they can make ATP. To a lesser extent hemoglobin binds to carbon dioxide to carry it away from your cells and to your lungs, to be expelled. Each red blood cell contains hundreds of millions of hemoglobin proteins. Figure 5.13 shows the structure of a healthy hemoglobin molecule.

**FIGURE 5.13**                                                    Hemoglobin molecule

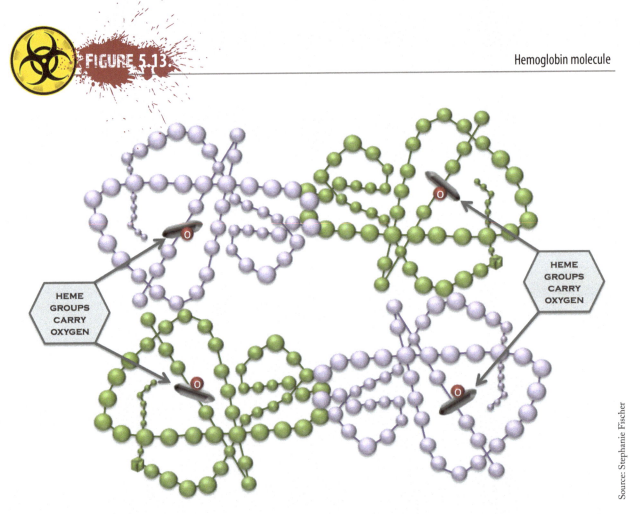

HEME GROUPS CARRY OXYGEN

HEME GROUPS CARRY OXYGEN

Source: Stephanie Fischer

Like any protein, hemoglobin is made up of a string of amino acids (and the order of the amino acids is coded for by DNA). On two of the corners there are glutamic acid amino acids. Glutamic acid is polar, and therefore likes to be associated with water in the cell. Because of this hydrophilic property of glutamic acid, the hemoglobin spreads out in the red blood cell and distributes evenly. This gives red blood cells their spherical shape.

There is a particular genetic mutation that can happen in the hemoglobin gene. A **single** DNA base change alters **one** amino acid, from a glutamic acid to a valine.

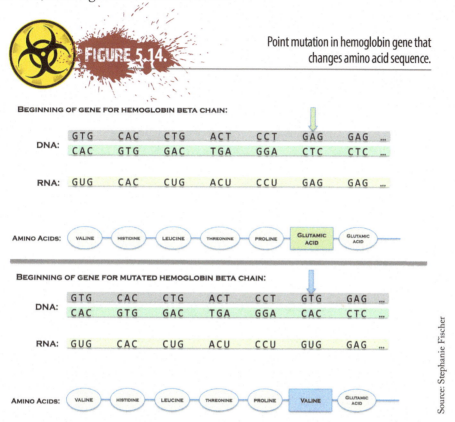

FIGURE 5.14.

Point mutation in hemoglobin gene that changes amino acid sequence.

BEGINNING OF GENE FOR HEMOGLOBIN BETA CHAIN:

DNA:
| GTG | CAC | CTG | ACT | CCT | GAG | GAG | ... |
| CAC | GTG | GAC | TGA | GGA | CTC | CTC | ... |

RNA:
| GUG | CAC | CUG | ACU | CCU | GAG | GAG | ... |

AMINO ACIDS: VALINE — HISTIDINE — LEUCINE — THREONINE — PROLINE — GLUTAMIC ACID — GLUTAMIC ACID

BEGINNING OF GENE FOR MUTATED HEMOGLOBIN BETA CHAIN:

DNA:
| GTG | CAC | CTG | ACT | CCT | GTG | GAG | ... |
| CAC | GTG | GAC | TGA | GGA | CAC | CTC | ... |

RNA:
| GUG | CAC | CUG | ACU | CCU | GUG | GAG | ... |

AMINO ACIDS: VALINE — HISTIDINE — LEUCINE — THREONINE — PROLINE — VALINE — GLUTAMIC ACID

Source: Stephanie Fischer

Figure 5.14 shows the mutation site. Unlike the hydrophilic glutamic acids, the valines all bind together forming a clumped or crystallized structure in the red blood cell (figure 5.15). This changes the shape of red blood cells into a shriveled, sickle shape, as shown in figure 5.16. (Note that the scale in this figure is exaggerated; there are hundreds of thousands of hemoglobin proteins in a single red blood cell.) This is an example of a *single* base change in the DNA causing a change in the amino acid sequence of a protein, and that changing the shape and function of the protein leads to a disease state.

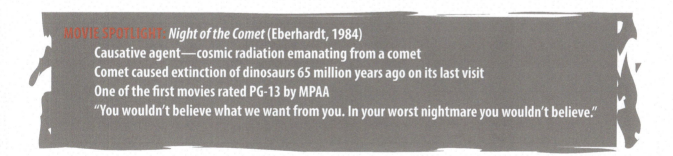

MOVIE SPOTLIGHT: *Night of the Comet* (Eberhardt, 1984)
Causative agent—cosmic radiation emanating from a comet
Comet caused extinction of dinosaurs 65 million years ago on its last visit
One of the first movies rated PG-13 by MPAA
"You wouldn't believe what we want from you. In your worst nightmare you wouldn't believe."

 **FIGURE 5.15.** Sickle-cell mutation changes glutamic acid to valine in beta amino acid chain.

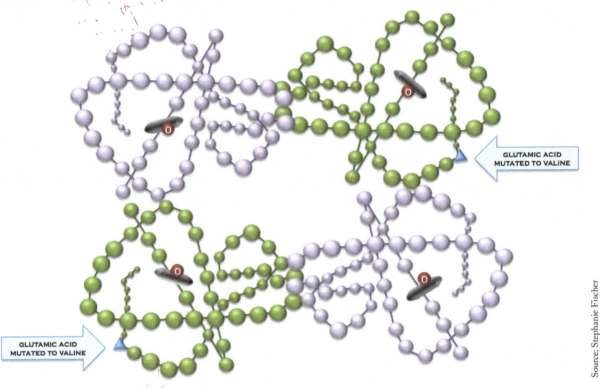

GLUTAMIC ACID
MUTATED TO VALINE

GLUTAMIC ACID
MUTATED TO VALINE

Source: Stephanie Fischer

 **FIGURE 5.16.** Distribution of hemoglobin in RBC changes due to mutation.

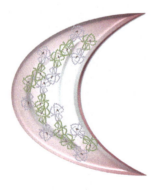

NORMAL HEMOGLOBIN
IN RED BLOOD CELL

MUTATED HEMOGLOBIN
FORCING RED BLOOD CELL
INTO SICKLE SHAPE

*NOTE THAT THIS SCALE IS EXAGGERATED

Source: Stephanie Fischer

Healthy red blood cells are 7 to 8 μm in diameter, and our smallest blood vessels, our capillaries, are not much wider. The circular-shaped healthy red blood cells stack like poker chips to pass single file through the capillary. The sickle-cell shapes are more likely to get stuck and cause a blockage in the blood vessels. Patients stricken with sickle-cell anemia are more prone to heart attacks and strokes as a result of these deformed molecules. It also causes excruciating pain and there is no cure, only treatment of symptoms.

Now, not every base change causes a change in the amino acid sequence (there is some redundancy in the code). In addition, not every change changes the function all that much, and not every change is detrimental to the organism.

Not all mutations are single base mutations, either. Some are duplications of whole sections of DNA code, some are deletions of whole sections of DNA code. Some flip sections of the code. Table 5.1 shows a brief description of many types of mutations.

**TABLE 5.1**                                        Common mutations

| Mutation | Resulting message |
|---|---|
| Original code | THE ZOMBIE IS COMING! |
| Original code with codons | THE ZOM BIE ISC OMI NG! |
| Single base change | THE ZUM BIE ISC OMI NG! |
| Reversing order | THE EIB MOZ ISC OMI NG! |
| Deletion | THE OMB IEI SCO MIN G!_ |
| Insertion | THE ZOO MBI EIS COM ING !__ |
| Large deletion | THE BIE ISC OMI NG! |
| Large insertion | THE BAD ZOM BIE ISC OMI NG! |
| Nonsense insertion | THE ZOM |

So, then the question for our hypothesis (that genetic mutation triggered by radiation causes our zombie outbreak) becomes a question of… is there a way to mutate one or two proteins that would result in zombieism?

Proteins do all the different jobs in your cells and organelles to keep everything working properly. As stated earlier, segments of DNA called genes are the instructions for how to make each protein. A mutation in a gene (sequence of DNA) can mess up the structure of a protein, which in turn means that protein *may* not do its job, which means an organelle may not do its job. If that organelle is not doing its job, then the cells can malfunction and the organism can have symptoms… a disease state caused by that malfunctioning protein.

This is pretty oversimplified, but it is how it works in general terms.

## TAY-SACHS DISEASE

One example of how a mutation in one protein can shut down an organelle is Tay-Sachs disease.

In Tay-Sachs, there is a mutation in one of the enzymes to break down a component of the cell membrane called a ganglioside. The enzyme responsible for this task is called hexosaminidase A. This enzyme is usually stored in an organelle called a lysosome, which you may recall is a recycling center of the cell. Because the enzyme does not function (wrong DNA bases, wrong amino acid sequence, wrong shape, no function!), the fatty membrane components (gangliosides) build up in the child's cells, especially the brain. The children begin to deteriorate mentally, then physically, and usually die by the age of 4.

In order for a child to inherit the disease, both parents must carry a mutation in the gene (even one copy of a functioning gene in one parents' strand of DNA is enough to keep a person healthy). If the DNA in the gene (instructions for this protein) from both parents is mutated, the child will not make any functional enzyme, and will have the disease. It is fairly rare, but does occur with greater frequency in populations that have been isolated at some point in history.

So… genetic mutations can definitely cause diseases.

## GUARDIAN OF THE GENOME: P53

Several proteins have been discovered that monitor newly made DNA for errors, and repair broken DNA. These proteins play a big role in preventing **cancer**, because they stop cells with mutations or errors from reproducing until those mistakes are fixed. (If the mistakes cannot be fixed, the proteins force the cell to undergo **apoptosis**, or a "quiet death.")

The **p53 protein** is one of those guardian proteins; it skims along new DNA strands, and if it finds a mismatch or error or break, it forces the cell to fix it or undergo apoptosis. It turns out that almost half of all tumors have been found to have a malfunctioning p53 protein (Levine et al.) due to a mutation in the **p53 gene** that codes for the protein; meaning this method of finding and forcing the repair of mutations is critical to preventing cancer. Other proteins have been found that similarly "guard" against mutations, such as the **BRCA protein** (a mutation in the gene that codes for the BRCA protein has been shown to increase a person's risk of breast cancer).

Elephants have a lower rate of cancer than humans, which has been linked to elephants having extra (20!) copies of the p53 gene, so that they have backup copies to make sure they have fully functioning p53 proteins (Padariya et al.).

Infection with the **Human Papilloma Virus (HPV)** can increase a person's risk of cancer, because the virus contains a gene for a protein (E6) that binds to p53, stopping it from working and tagging it for destruction by the cells it infects. HPV integrates this viral gene for E6 into the cells it infects, and those cells then degrade their own p53 proteins. Over time, these cells don't have enough p53 to correct mistakes made during DNA replication, and the cells accumulate mutations with a greater risk of becoming cancerous.

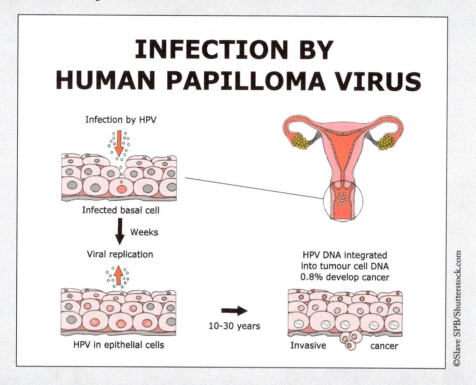

# INFECTION BY HUMAN PAPILLOMA VIRUS

Infection by HPV

Infected basal cell

Weeks

Viral replication

HPV in epithelial cells

10-30 years

HPV DNA integrated into tumour cell DNA 0.8% develop cancer

Invasive          cancer

©Slave SPB/Shutterstock.com

Levine AJ, Oren M. The first 30 years of p53: growing ever more complex. Nat Rev Cancer. 2009 Oct;9(10):749-58. doi: 10.1038/nrc2723. PMID: 19776744; PMCID: PMC2771725.

Padariya M, Jooste M-L, Hupp T, Fåhraeus R, Vojtesek B, Vollrath F, Kalathiya U, Karakostis K. The elephant evolved p53 isoforms that escape MDM2-mediated repression and cancer. *Molecul Biol. Evol.* 2022 July;39(7):msac149, https://doi.org/10.1093/molbev/msac149

## BUT CAN RADIATION CAUSE ZOMBIES?

So, let's get back to how zombies might be created by radiation… we've said that mutations can lead to disease symptoms… can radiation lead to zombie symptoms directly? Or would it have to work through gene mutations?

Radiation has some immediate effects: it is a high-energy source so it can literally burn its victims to death if they are close enough. For those farther away, inhaled radiation can cause searing of the lungs leading to the lungs becoming inflamed and filling with fluid until the victims drown in their own immune system fluid.

But if you are farther away from the radiation source, death comes more slowly and more invisibly.

Radiation is a type of energy... and it is a type of energy that can penetrate through skin and membranes and basically bang into DNA, causing mutations. Figure 5.17 shows the different wavelengths of radiation on the electromagnetic spectrum.

So, a single mutation (like Tay-Sachs disease or sickle-cell anemia) probably can't cause the kind of symptoms we see with zombies, any more than losing a single neurotransmitter can—the brain is just too complex to get zombie symptoms from one protein malfunctioning. But... what about multiple proteins failing all at once? It's conceivable with radiation poisoning.

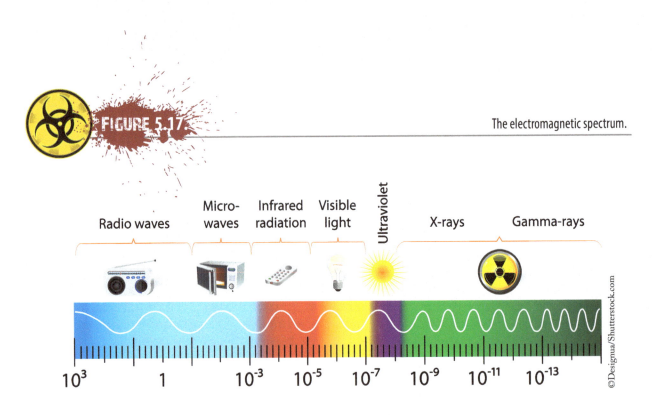

**FIGURE 5.17** The electromagnetic spectrum.

Radiation has enough energy to break up numerous sections of DNA, or cause several DNA mutations at once. Because it can cause all this DNA damage at once, it can lead to the development of cancerous tumors. As cells mature, they **differentiate**, or become specialized. Remember, all the cells in your body have the same set of DNA, or instructions for building all the proteins. But they don't all produce every possible protein; as they mature, neurons produce neuron proteins that they need, while red blood cells produce just the proteins they need. This is a way of specializing in certain cell functions, and there are special controls that work to shut down some unneeded genes to allow this specialization. **Tumor** cells have mutations that reverse this specialization, and they begin to reproduce in an out-of-control way. They can even detach themselves and move through the body, setting up new colonies or tumors in other places (we call this movement "**metastasizing**"). So low-dose radiation exposure can lead to random DNA mutations that encourage tumor formation and the development of cancer.

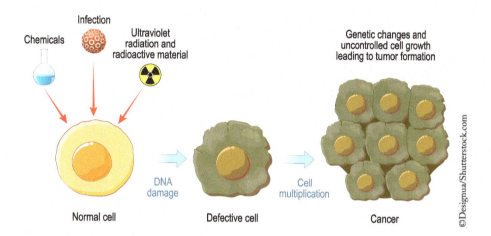

Normal cell     Defective cell     Cancer

©Designua/Shutterstock.com

Interestingly, kids that are exposed to radiation are more likely to get thyroid cancers and skin cancers, while adults exposed to radiation are more likely to get lung and blood cancers. Most importantly for our zombie stories, these cancers are not the same for every person, and they are not contagious cancers, making zombies-caused-by-radiation even more unlikely.

Eventually organs begin to fail as the mutated proteins cannot work properly, leading to a slow, agonizing death. Additionally, incidents like Hiroshima, Nagasaki, and Chernobyl show that radiation can cause mutations in wide swaths of the population at once.

## TALKING ABOUT RADIATION EXPOSURE

In a recent study of identical twin astronauts Mark and Scott Kelly, scientists monitored each twin's DNA changes while Mark stayed on Earth and Scott spent 340 days on the International Space Station. Scientists saw changes in Scott's immune system function, cognitive performance, and increased DNA damage, in addition to other changes. Scientists hypothesize that these changes are due to the physical stresses of weightlessness and living on the ISS, as well as due to heightened exposure to radiation from the sun and space (Earth's magnetic field protects us Earth-dwellers from most of this radiation). This was a great scientific study of low-levels of radiation resulting in a multitude of small changes that could add up to increased infection rates, changes in the immune system, changes in brain function, and many small mutations in a person's DNA.

Credit: NASA

Now, actually, this can come pretty close to zombie stories, right? It would take time, but the survivors of such an exposure to a radiation burst might survive for awhile and then slowly begin to show symptoms. Before you head to the radiation-proof bomb shelter, however, there are several problems with the radiation-as-the-cause-of-zombies hypothesis. First, even though one strong burst of radiation can damage thousands of people, each mutation in each person would be different. It's simply not possible for multiple victims to show the exact same mutation-based zombie-like symptoms. Second, even if a victim gets zombie-like traits, mutation diseases aren't contagious. So one or a handful of noncontagious zombies aren't exactly going to cause an outbreak. So radioactive zombies? Nope. But radioactive teenagers with the proportionate strength and speed of a spider? *Much* more plausible.

## CANCER TREATMENT

Interestingly, the treatment of cancer cells involves causing DNA damage (even though cancer itself is often caused by DNA damage); this time, by causing DNA damage to the tumor cells themselves. The idea is, since tumor cells are reproducing much faster than most body cells, **chemotherapy** drugs and radiation treatments that damage DNA can hopefully be used to cause enough DNA damage to cause cancer cell death. This is why chemotherapeutic drugs and radiation treatments cause the death of other rapidly dividing body cells, like hair cells and white blood cells, leaving cancer patients **immuno-compromised** during their treatment.

Another type of treatment targets the food supply of tumor cells. When tumor cells form, they send signals out for the body to build more blood vessels to feed them, this is called **angiogenesis** (angio = blood, genesis = creation). This makes the tumor cells almost like zombies, using the body for food to reproduce and spread. Some cancer drugs target this process by blocking these signals.

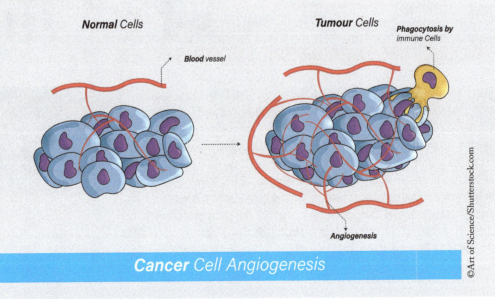

*Normal* Cells   *Tumour* Cells

**Phagocytosis by** immune Cells

**Blood** vessel

Angiogenesis

©Art of Science/Shutterstock.com

**Cancer** Cell Angiogenesis

That isn't to say radiation can't still be an accomplice to a zombie outbreak. As it turns out, viruses and bacteria contain DNA and RNA also. So some of the more creative zombie stories involve not simply a dangerous virus or bacterium, but a virus or bacterium that has been mutated by radiation, leading to a completely new pathogen that causes zombie symptoms. But that is really a story for the next chapter.

**FIGURE 5.18.** Spins a web, any size, catches thieves just like flies.

©Anton_Ivanov/Shutterstock.com

# CHAPTER 5 QUESTIONS/WORDSTEMS

1. Transcription occurs in the
   a. Nucleus.
   b. Ribosomes of the Rough Endoplasmic Reticulum.
   c. Mitochondrion.
   d. Cell membrane.
   e. Smooth Endoplasmic Reticulum.

2. The monomers of DNA and RNA are
   a. amino acids.
   b. monosaccharides.
   c. nucleotides.
   d. fatty acids.
   e. nucleic acids.

3. Which of the following statements regarding DNA is *false*?
   a. DNA uses the nitrogenous base uracil.
   b. DNA is a nucleic acid.
   c. One DNA molecule can include four different nucleotides in its structure.
   d. DNA uses the nitrogenous base thymine.
   e. DNA molecules have a sugar-phosphate backbone.

4. Which of the following options best depicts the flow of information when a gene directs the synthesis of a cellular component (transcription and translation)?
   a. DNA → mRNA → protein
   b. mRNA → DNA → protein
   c. protein → mRNA → DNA
   d. mRNA → protein → DNA
   e. DNA → protein → mRNA

5. Experiments have demonstrated that the "words" of the genetic code (that are transcribed into codons) are
   a. single nucleotides.
   b. two-nucleotide sequences.
   c. three-nucleotide sequences.
   d. nucleotide sequences of various lengths.
   e. enzymes.

6. Which of the following takes place during translation?
   a. translate info from mRNA to a protein.
   b. translate info from DNA to mRNA.
   c. translate info from DNA to tRNA.
   d. translate info from proteins to enzymes.
   e. DNA replication.

| WORDSTEMS | |
|---|---|
| muta- | change |
| -gen | give rise to |
| carcin- | crab; cancer |
| radi- | ray; spoke |
| trans- | to cross over, change |
| -late | from "relate", language |
| -scribe | write |
| hemo- | blood |
| oblig- | to bind to |
| -globin | specific protein |
| nucl- | nut; kernal |

# CHAPTER 5 — WORKSHEET

## PROTEINS AND MUTATIONS (GROUP PROJECT)

### OBJECTIVES

1. To introduce the steps cells take during protein synthesis
2. To study the effects of mutation on protein synthesis

### INTRODUCTION

We already discussed the importance of proteins and enzymes in chapters 3 and 4. But to fully understand the flow of genetic information (DNA to final protein) we have to delve into the cellular metabolic process known as protein synthesis. Today, you and your group will be constructing a protein from scratch using just the information from DNA.

### PROTEIN SYNTHESIS

### MATERIALS

Colored pencils or highlighters

### PROCEDURE

1. Working in groups of three you will follow the flow of genetic information from DNA to RNA to proteins. Start with the following DNA strand containing our gene. Identify which is the template strand and which is the non-template (copy) strand.

2. How can you tell?

| DNA | | T A T A A A A G C C T C A T T A C C C A G C A G G A A A A T A C A T T T C G C A A T C |
|-----|--|--------------------------------------------------------------------------------------|
| | | A T A T T T T C G G A G T A A T G G G T C G T C C T T T T A T G T A A A G C G T T A G |

3. Highlight or color in the TATA box. This identifies the start of the gene, as well as the template strand.

4. Now build the corresponding mRNA from the template strand.

☐☐☐☐☐☐☐☐☐☐☐☐☐☐☐☐☐☐☐☐☐☐☐☐☐☐☐☐☐☐☐☐☐☐☐☐☐☐☐☐☐☐☐☐☐☐☐☐☐☐☐☐☐☐

5. What is the process of making mRNA from DNA called?

6. Where in the cell does this process take place?

Second Base

| First Base | | U | C | A | G | | Third Base |
|---|---|---|---|---|---|---|---|
| U | | UUU phenylalanine | UCU serine | UAU tyrosine | UGU cysteine | U | |
| | | UUC phenylalanine | UCC serine | UAC tyrosine | UGC cysteine | C | |
| | | UUA leucine | UCA serine | UAA stop | UGA stop | A | |
| | | UUG leucine | UCG serine | UAG stop | UGG tryptophan | G | |
| C | | CUU leucine | CCU proline | CAU histidine | CGU arginine | U | |
| | | CUC leucine | CCC proline | CAC histidine | CGC arginine | C | |
| | | CUA leucine | CCA proline | CAA glutamine | CGA arginine | A | |
| | | CUG leucine | CCG proline | CAG glutamine | CGG arginine | G | |
| A | | AUU isoleucine | ACU threonine | AAU asparagine | AGU serine | U | |
| | | AUC isoleucine | ACC threonine | AAC asparagine | AGC serine | C | |
| | | AUA isoleucine | ACA threonine | AAA lysine | AGA arginine | A | |
| | | AUG(start) methionine | ACG threonine | AAG lysine | AGG arginine | G | |
| G | | GUU valine | GCU alanine | GAU aspartate | GGU glycine | U | |
| | | GUC valine | GCC alanine | GAC aspartate | GGC glycine | C | |
| | | GUA valine | GCA alanine | GAA glutamate | GGA glycine | A | |
| | | GUG valine | GCG alanine | GAG glutamate | GGG glycine | G | |

© Kendall Hunt Publishing Company

7. Now refer to the genetic code provided. This is a translation from codons on mRNA to amino acids. You will use this table to make your final protein.

8.  What is the process of making the final protein from mRNA called?




9.  Where in the cell does this process take place?




10.  Find the codon AUG. What amino acid does this code for?




11.  What else is special about this particular codon?




12.  Starting from left to right, search for the first appearance of the codon AUG. Highlight or color in the codon in green. Translation doesn't start until the ribosome finds the AUG codon. Now highlight or color in the codons that proceed the start codon, all with different colors (but don't use red). When does the ribosome know when to stop translating?




13.  Highlight the stop codon in red.




14.  Now that we've identified all of the codons, write the corresponding amino acids below. Highlight or color in the amino acids to match the colors on the codon.


| | | | | | |
|---|---|---|---|---|---|
| | | | | | |


15.  This is your finished (albeit very small) protein. How large is a typical protein?




16.  Now we'll take the mRNA you constructed to find out the effects of mutations.

## Mutations

### Procedure

17. Now that you have your finished protein, let's see what changes to the DNA can do to that protein. We will be focusing on *point mutations*, or mutations that affect only one nucleotide in the gene sequence.

18. Look back to your DNA and find the sequence GGA. You should have already determined that this transcribed into the codon CCU. What codon did this translate into?

19. Now let's see what happens when this mutation occurs. GGA → GGG

20. Write the corresponding codon, followed by the resulting amino acid.

21. Since this mutation does not change the final amino acid, we call it a *silent mutation*. A silent mutation would have no effect on the final protein.

22. Go back to your DNA to the same sequence, GGA. Now let's see what happens when this mutation occurs. GGA → AGA

23. Write the corresponding codon, followed by the resulting amino acid.

24. Since this mutation changes the amino acid, we call it a *missense mutation*. Even though this would be just one amino acid out of a few thousand in a typical protein, it can still prove damaging. What genetic disease did we talk about that was caused by the changing of a single amino acid?

25. Let's do one last mutation. Instead of replacing one nucleotide, however, we're going to show a mutation that dislodges a nucleotide. GGA → GGA

26. This is called a frameshift mutation, as all other nucleotides slide down. Show how this can disrupt not just one codon, but *all remaining codons* as well.

# CHAPTER 6

## THE ZOMBIE PLAGUE

**(OR... WASH YOUR HANDS, COVER YOUR MOUTH WHEN YOU COUGH, AND RESIST YOUR PRIMAL AMOROUS URGES)**

So yes, voodoo curses gave us the original zombie, and the fear of a nuclear holocaust in the 1960s gave us *Night of the Living Dead* and popularized the undead. But this is the new millennium. Those seriously looking to prevent the zombie apocalypse turn their heads toward microorganisms.

Microorganisms can include prokaryotic bacteria, eukaryotes like fungi, protozoans and microscopic animals, or even nonliving parasites like viruses and prions. However, not all microorganisms cause disease. In point of fact, the majority of microorganisms are beneficial to life on this planet or at least harmless. Many are actually essential to our survival.

In this chapter we will be focusing on human pathogens, which are disease-causing microorganisms. These organisms parasitize you either by feeding off you or stealing your energy, doing you bodily harm in the process.

# BACTERIAL CELLS: PROKARYOTES

What is the dominant form of life on this planet? You? Ha! Whatever gave you that idea? Because we have opposable thumbs? Because we invented technology like the shake weight? Maybe because we can create and appreciate art like the Macarena? Ohhhh, the hubris. Everything else on the planet is fighting for a distant second behind bacteria. It's more accurate to say this is the age of the bacterium. Come to think of it, since the dawn of time *every age* has been the age of bacteria.

There's not a habitat on the planet that you won't find bacteria. Too cold for eukaryotes? Too hot? Too acidic too basic too dry too salty too… bacteria scoff at us fragile eukaryotes. Bacteria outnumber every eukaryote combined with respect to both number of cells and diversity. As a matter of fact, there are 10 times more bacterial cells inside you right now than there are human cells. Essentially, there are more bacteria in you than there is *you* in you.

 **FIGURE 6.1.** ———————————————————— Anatomy of a prokaryotic bacterium.

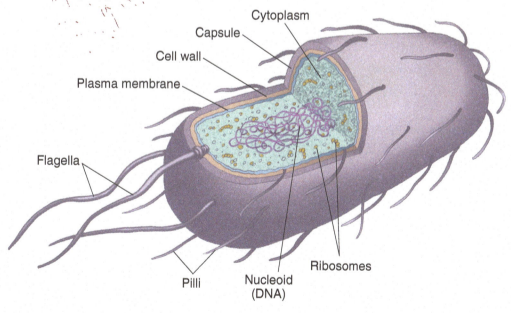

Cytoplasm
Capsule
Cell wall
Plasma membrane
Flagella
Ribosomes
Pilli
Nucleoid
(DNA)

There are numerous differences between our eukaryotic cells and prokaryotic bacteria, but the two main distinctions are size and complexity. Turns out these two traits are intertwined.

First, size. The typical eukaryotic (animal, plant, fungus, algae, protozoan) is between 10 and 100 μm (millionths of a meter) in diameter. A prokaryote, however, is only 1 μm in diameter. Essentially, a bacterium is up to 100 times smaller than your typical human cell. Figure 6.2 shows the difference in size between the buccal (epithelial) cells lining your oral cavity and the bacteria found there. The smaller blue structures are the bacteria while the larger lighter cells are the human cell. The darker stained structure inside of the eukaryote is the nucleus.

The larger size of the eukaryotes is not an advantage. In fact, it's a distinct disadvantage. That's because as size increases, surface to volume (S/V) ratio decreases exponentially. That's the amount of surface area of the cell compared to its volume. Let's take three spherical cells for example: a 1 μm prokaryote, a small 10 μm eukaryote, and a larger 100 μm eukaryote like your buccal cell. Table 6.1 shows the side by side comparison.

FIGURE 6.2.

Oral scraping showing buccal cells and bacteria (400x TM, stained with methylene blue).

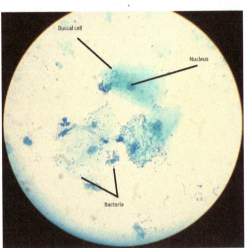

Source: Christopher Green

**TABLE 6.1**

Comparison of cells by size.

|  | prokaryote | small eukaryote | large eukaryote |
|---|---|---|---|
| **Diameter (μm)** | 1 | 10 | 100 |
| **Surface area (μm²)** | 13 | 1,256 | 125,600 |
| **Volume (μm³)** | 4 | 4,190 | 4,190,000 |
| **S/V ratio** | 3 | 0.3 | 0.03 |

So why is the S/V ratio so important? The cell membrane is where everything enters and leaves the cell. The more surface area with respect to the volume, the faster the cell can grow. Picture this analogy. Take two identical ice cubes out of your freezer. Same size and temperature. Now take one of those cubes and crush it. Which one melts faster in your (still not real) Zombie-Cola? The crushed one. Much faster. Why? Greater surface to volume ratio. So bacteria can grow and reproduce so much faster than eukaryotes. We humans take 20 years to go from generation to generation. Prokaryotes can do it *in 20 minutes* (via binary fission).

Secondly, bacterial cells are a lot simpler than our cells. Notice from the image of the oral scraping that there is no nucleus in a bacterial cell. Rather, their DNA is coiled in a central region of the cell, with no phospholipid membrane separating the DNA from the rest of the cell. There are also no membrane-bound organelles within the bacterial cell: no ER, no Golgi, no mitochondria. However, there are lots of ribosomes, which do the same job they do in our cells: they "read" the information in mRNA and make proteins from that information.

So bacteria are smaller and simpler with only two goals: eat and reproduce. They do so with remarkable speed. Give one bacterium plenty of food and by the next morning you have billions.

## HOW DO BACTERIA CAUSE DISEASE?

Pathogenic (disease-causing) bacteria can cause symptoms in a host (like a human) in several different ways. Often they simply do what they do best: eat and reproduce. Well, often what they're eating… is you. *Streptococcus pyogenes* eats the epithelial tissue at the back of your throat and you suffer from strep throat (bacterial pharyngitis). *Staphylococcus aureus* eats into a wound on your arm and you suffer from a staph infection (cellulitis). These pathogens are stealing energy (food) from the host which can cause problems with the host cells being able to function. Many times rapid bacterial growth can physically displace so many human cells so that the organ the bacteria are infecting can no longer function.

More commonly, the bacteria produce toxins (chemicals that interfere with cell or protein function). These toxins can destroy cells. For example, one toxin that is commonly found in Methicillin Resistant *Staphylococcus aureus* (MRSA) is able to destroy white blood cells. *Clostridium botulinum* is the causative agent of the potentially fatal disease botulism; it secretes a neurotoxin called botulin which inhibits acetylcholine release (this keeps muscles from contracting). You might want your diaphragm or your heart to keep contracting… (and yes we inject the same stuff and call it BoTox… man, we're just asking for an apocalypse).

But a large part of how bacteria ultimately kill patients is by overwhelming the immune system to the point where the patient goes into shock and dies. So much of the body's energy is devoted to trying to fight off the pathogen (and especially if the pathogen reaches the brain), that the body is overwhelmed by demands. Often if the lungs become infected or inflamed the patient can't take in oxygen any more. Bacterial toxins can destroy kidney cells, causing self-made toxins to build up in the blood and destroy the patient's brain. If the infection reaches the brain itself it can destroy the control center (hypothalamus and brain stem) of the body. The patient loses the ability to regulate heartbeat, breathing, and temperature.

### LEPROSY

Could a bacterial disease lead to zombie symptoms? Right now, all antibiotic resistant bacterial diseases are pretty frightening and a real danger to public well-being. But most of them kill rapidly and their victims don't resemble zombies at all. So, infections can cause fevers, brain inflammation, confusion, pneumonia, organ failure, and death? Yes. But zombie symptoms? Not quite. There are, however, bacterial diseases that at least have the outward appearances similar to zombieism.

The disease leprosy (also known as Hansen's disease) is caused by the bacterium *Mycobacterium leprae*. It can cause nerve damage which eventually leads to skin deformities and circulation problems. That might look a lot like the typical movie zombie. In fact, leprosy is one of the oldest recorded diseases.

FIGURE 6.3.

Hands of a leprosy patient.

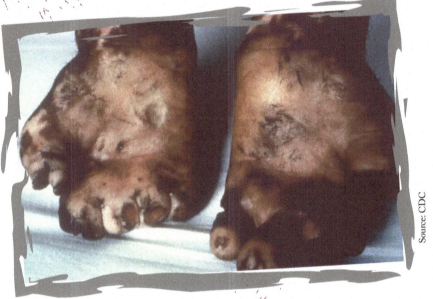

Source: CDC

Leprosy was written about in ancient texts across the world, including China, India, Greece, Egypt, and the Middle East. Earliest writings date back to more than 4000 BCE.

But leprosy doesn't cause brain problems or lack of rational thought, although it can cause lack of sensation and nerve conduction from the limbs. So, not a mindless zombie, but definitely similar to our zombie appearance, gait, and insensitivity to pain. Because leprosy is such an old disease and such a big

FIGURE 6.4.

Leper in Martinique c. 1902.

Source: Library of Congress

part of the human experience throughout history, it is likely that the physical symptoms of leprosy did factor in to our mythological creation of zombies.

We can now cure early cases of leprosy, and treat even the more advanced cases (although not always cure the most advanced cases), with a regimen of several different antibiotics. We'll discuss antibiotics later on. So far, there is not an antibiotic resistant form of leprosy. Leprosy can take up to 20 years to incubate and begin to show symptoms (first lesion). It usually takes 3 to 10 years to show first symptoms. It is a very slow growing bacterium.

Could a mutated form of the leprosy bacteria cause zombies? The bacteria that cause leprosy do not cause changes in brain function, so patients retain full rationality. But some bacteria related to the leprosy bacteria could potentially cause the zombie-like appearance. It is important, though, to remember how very slowly this disease develops. While this was a very frightening disease to the ancients, a slowly developing disease that can be stopped by antibiotics doesn't really fit the rapidly spreading zombie profile as is… but with the right mutations? Who knows?

## THE BLACK DEATH

The bacterial pathogen *Yersinia pestis* was responsible for one of the worst pandemics in human history, responsible for the death of up to 60% of the European population between 1346 and 1353. It is the causative agent of several forms of plague, including pneumonic plague in the lungs and bubonic plague in the lymph nodes. Victims of bubonic plague suffer from swollen buboes (infected lymph nodes) in the groin, neck, and armpits which ooze pus and blood when ruptured. Other symptoms include acute fever and vomiting of blood.

**FIGURE 6.5.**
                                                *The Triumph of Death* by Pieter Bruegel the Elder c. 1562.

Prado, Madrid, Spain/Bridgeman Images

One of the most insidious traits of *Yersinia pestis* is the numerous ways it can be transmitted. It can be spread by droplets (sneezing or coughing) or direct contact (touching or sexual contact with an infected person). It can be spread by air, via fomites (inanimate objects), from contaminated food and water, and via vectors (animal carriers). Essentially the pathogen can be transmitted by every route known. We'll discuss modes of transmission in chapter 8.

Like leprosy, the plague gives zombie-like appearance to its victims but does not affect brain function. As such it is also not a viable candidate for a zombie outbreak barring significant mutations. However, its multiple modes of transmission can give us pause… a disease like the Hollywood forms of zombieism that only spreads via bites is far less dangerous than one with multiple forms of transmission like *Yersinia pestis*.

All told, however, there isn't a realistic bacterial pathogen that may cause anything close to what we think of as zombie-like symptoms. As stated earlier, bacteria simply eat and divide, eat and divide, of singular purpose for survival. So while bacteria may be the most successful life forms on the planet, we may have to look elsewhere in our pursuit for potential causes for the zombie apocalypse.

# VIRUSES

Viruses are all nonliving parasites that take over your cells from the inside and cause disease (nonliving monsters? …sounds like zombie microorganisms). They have no cellular structure and no metabolism. They are made of simple DNA or RNA (not both, as in living cells) surrounded by a protein capsid. Some contain an envelope made of a phospholipid bilayer outside the capsid, but that's it. No organelles, no ribosomes, no enzymatic activity outside of their host cell. The envelope is a part of the host's own cell membrane. It acts as a camouflage and allows the virus to attach to the next host (how badass is that… wrapping yourself in your victims flesh?). These envelopes also have embedded in them protein spikes (glycoproteins) for attachment to hosts. Figure 6.6 shows the structures of select viruses. Viruses are significantly smaller than even bacteria; their size is measured in nm (billionths of a meter). The average virus is 1/100th the diameter of a bacterium or smaller and up to 1/10,000th the size of our human cells. But don't be fooled… these submicroscopic critters are the cause of innumerable diseases that have plagued mankind throughout history.

Viruses are obligate intracellular parasites. This means that in order to replicate (we can't really say reproduce as they're not alive), viruses have to infect a host cell from the inside. That host might be a human cell, animal cell, plant cell, or a bacterial cell depending on the virus and the type of proteins it has on its surface. While only a small percentage of bacteria are pathogens (the majority are harmless or even beneficial to us)

FIGURE 6.6.

Disparate viral shapes.

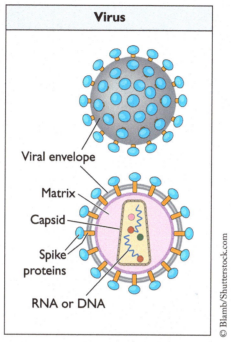

you can't say the same about viruses. *All* viruses are pathogens. They might not all be pathogens to you, but they parasitize *something* on this planet.

Since viruses have no metabolism on their own, the virus hijacks the cell's metabolism.

## VIRAL REPLICATION

The process by which the virus takes over the cell is referred to as the lytic cycle. The steps involved are summarized here.

*Attachment*: Most viruses have a very narrow host range; in other words most viruses can only attach to one type of organism's cells. Viruses target host cells based on what spikes they have on their surface. For the flu virus, for example, one type of protein spike on its surface is hemagglutinin, which binds to sialic acid containing proteins on the surface of human cells. Some human cells contain a lot of sialic acids on their surfaces, especially those that need to stay moist… like in your respiratory tract.

*Penetration*: In order to take over a cell the virus must inject its DNA or RNA into it. The protein spikes pull the virus close to the host cell and then the membranes fuse together, allowing the virus to release its nucleic acid into the host cell.

*Synthesis*: Once the nucleic acid gets inside the cell the enzymes already in the host start reading the viral nucleic acids. They start following instructions from the virus instead of its own cellular DNA. The virus instructs the cell to start making viral parts; viral nucleic acids are replicated, the capsid is formed, and spikes are embedded into the cell membrane.

*Assembly*: The virus then instructs the cell to assemble the viral parts, assembling the capsid around the viral nucleic acid. The host cell can make thousands of viruses before the stresses start killing the cell.

*Release*: The newly created viruses escape out of the host cell; on the way out they're wrapped in the cell membrane of the dying host (complete with built in spikes). The virus now complete with envelope goes off and finds another host to parasitize. As more and more viruses escape the cell shrivels (remember it's losing membrane with every virus) until it can no longer maintain integrity and spills open.

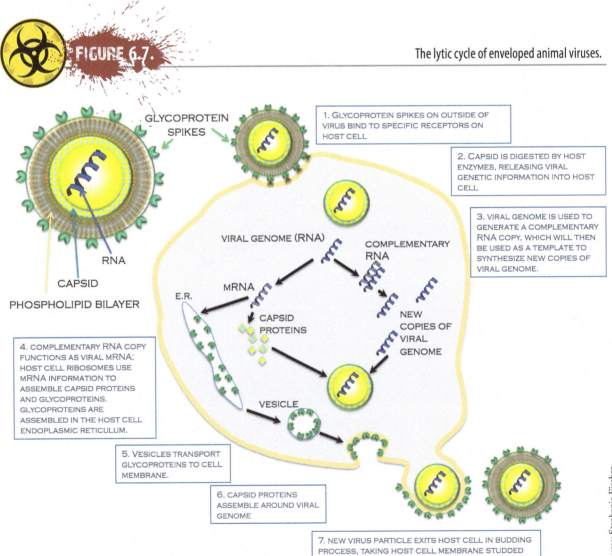

**FIGURE 6.7.** The lytic cycle of enveloped animal viruses.

GLYCOPROTEIN SPIKES

RNA

CAPSID

PHOSPHOLIPID BILAYER

VIRAL GENOME (RNA)

COMPLEMENTARY RNA

MRNA

E.R.

CAPSID PROTEINS

NEW COPIES OF VIRAL GENOME

VESICLE

1. GLYCOPROTEIN SPIKES ON OUTSIDE OF VIRUS BIND TO SPECIFIC RECEPTORS ON HOST CELL

2. CAPSID IS DIGESTED BY HOST ENZYMES, RELEASING VIRAL GENETIC INFORMATION INTO HOST CELL

3. VIRAL GENOME IS USED TO GENERATE A COMPLEMENTARY RNA COPY, WHICH WILL THEN BE USED AS A TEMPLATE TO SYNTHESIZE NEW COPIES OF VIRAL GENOME.

4. COMPLEMENTARY RNA COPY FUNCTIONS AS VIRAL MRNA; HOST CELL RIBOSOMES USE MRNA INFORMATION TO ASSEMBLE CAPSID PROTEINS AND GLYCOPROTEINS. GLYCOPROTEINS ARE ASSEMBLED IN THE HOST CELL ENDOPLASMIC RETICULUM.

5. VESICLES TRANSPORT GLYCOPROTEINS TO CELL MEMBRANE.

6. CAPSID PROTEINS ASSEMBLE AROUND VIRAL GENOME

7. NEW VIRUS PARTICLE EXITS HOST CELL IN BUDDING PROCESS, TAKING HOST CELL MEMBRANE STUDDED WITH VIRAL GLYCOPROTEINS.

The real question, however, is whether or not a theoretical virus is more capable than bacteria of causing a zombie outbreak. As it turns out, there are viruses that exist today that cause some zombie-like symptoms. Maybe just a little push (like from radiation-induced mutation to the viral genes?) and… blammo! Worldwide zombie party.

## RABIES

Rabies (acute encephalitis) is a viral disease with symptoms such as headache, fever, pain, uncontrolled excitement, mania, irritability, violence, and salivation (you know, like the new, popular fast zombie!). The aptly named Rabies virus (the name stems from the Latin word for

 **FIGURE 6.8.**

Rabies virus (SEM image).

Source: CDC/ Dr. Fred. A. Murphy

"raving") is a bloodborne pathogen spread by bites (again, zombie??!) and is 100% fatal if not treated promptly.

MOVIE SPOTLIGHT: *[REC]* (Balaguero, 2007)
Causative agent —mutated rabies virus
Remade for US audience as Quarantine (2008)
Ending suggests mutated enzyme, which would be characterized as a prion, not a virus
"There are incredible security measures in place. We know nothing. They haven't told us a thing. We saw Special Forces, health inspectors wearing suits and masks, and it's not very comforting."

**FIGURE 6.9.**

Rabies patient in restraints, 1959.

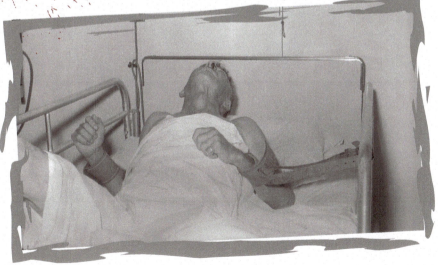

Source: CDC

The Rabies virus is also unlike most viruses in that it is a zoonosis (plural, zoonoses), meaning it can spread from animal to human. As stated earlier, most viruses can only parasitize one type of organism, meaning what gets you sick can't get your dog or cat sick. Well, zoonoses are the exception. And when viruses jump species, they often make the news (think avian, swine flu).

## EBOLA

In the summer of 2014 it seemed you couldn't turn on the TV without someone warning you about the dangers of Ebola. A viral disease with a high mortality rate, Ebola had previously only caused outbreaks in rural settings. Now, it had emerged in an urban city with international travelers. Many media outlets emphasized what could go wrong, rather than the carefully thought-out protocols for what to do correctly. Many media and social network sources were spinning the hype and not reporting the science, in order to increase panic and sell the story. While many in developed countries worried about Ebola causing an epidemic in their home cities, the reality was that with robust infection control protocols such as engineering controls and PPE (discussed further in chapter 7), the spread of Ebola in the United States was limited to two healthcare professionals. Both of these people survived, thanks in large measure to the excellent supportive care they received to keep their organ systems functioning while their immune systems mounted an antibody response to the virus. In the outbreak centers in Africa, though, extreme poverty, lack of clean water, and little or nonexistent healthcare turned controllable outbreaks into a deadly epidemic. By the end of the epidemic there were 28,000 cases resulting in 11,000 deaths. In some locations the mortality rate was as high as 70%.

**FIGURE 6.10**

Ebola patient quarantined.

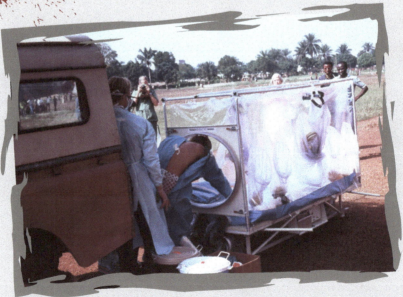

Source: CDC/ Dr. Lyle Conrad

So why isn't a zoonotic viral disease with a 100% fatality rate on everyone's radar? Why no urgent governmental warning against the dangers of rabid animals? Well, we do have a vaccine and antiviral treatment to halt the spread of infection (see chapter 7). But more importantly, it's how it spreads (transmission, chapter 8). Specifically, *mammalian bites are terribly inefficient* at spreading disease. Likely everyone reading this book has been nibbled or bitten numerous times by both dogs and cats in their lifetime. If bites guaranteed getting a disease you'd have CDC warnings on every kitten and puppy. In fact, take the rather infamous virus, rabies. Rabies viruses travel slowly up axons to the brain, where they infect and cause symptoms of agitation, pain, and paralysis (particularly of swallowing muscles). Once a person starts showing symptoms, death is certain. But luckily, we have a vaccine that helps block the virus from reaching the brain, if you can get it soon enough (before the virus reaches the brain). And, despite what you see in zombie stories, there are ZERO recorded cases of rabies passing from human to human via bite (but there have been accidental transmissions when an infected organ is transplanted). Once rabies reaches the brain, it is a horrifying illness, but rabies-infected people don't lose their ability to think rationally.

**BOOK SPOTLIGHT:** *World War Z: An Oral History of the Zombie War* (Brooks, 2006)
Follow up to Max Brooks' *The Zombie Survival Guide* (2003).
Tells the history of the zombie plague from beginning to end from the eyes of individual accounts narrated to an agent of the United Nations.
The film *World Ward Z* (Forster, 2013), loosely based on the novel, became the highest grossing zombie movie of all time.

**FIGURE 6.11.**

Yeah, this is a real news article.

The New Dawn (Monrovia) »                                                                                   24 SEPTEMBER 2014

### Liberia: Dead Ebola Patients Resurrect?

© Chaikom/Shutterstock.com

By Franklin Doloquee

Two Ebola patients, who died of the virus in separate communities in Nimba County have reportedly resurrected in the county. The victims, both females, believed to be in their 60s and 40s respectively, died of the Ebola virus recently in Hope Village Community and the Catholic Community in Ganta, Nimba.

But to the amazement of residents and onlookers on Monday, the deceased reportedly regained life in total disbelief. The New Dawn Nimba County correspondent said the late Dorris Quoi of Hope Village Community and the second victim only identified as Ma Kebeh, said to be in her late 60s, were about to be taken for burial when they resurrected.

Ma Kebeh had reportedly been in door for two nights without food and medication before her alleged death. Nimba County has had bizarre news of Ebola cases with a native doctor from the county, who claimed that he could cure infected victims, dying of the virus himself last week.

News of the resurrection of the two victims has reportedly created panic in residents of Hope Village Community and Ganta at large, with some citizens describing Dorris Quoi as a ghost, who shouldn't live among them. Since the Ebola outbreak in Nimba County, this is the first incident of dead victims resurrecting.

So yes, there are viruses that infect the brain; viruses could conceivably be responsible for a zombie-like outbreak. But perhaps there are other organisms even more likely to cause zombie-like symptoms... say organisms that could actually control your very brain. Now isn't that a scary thought?

## EUKARYOTIC BRAIN PARASITES

Parasitism is a type of symbiotic relationship where one organism (parasite) lives by harming another organism (the host). Most often, the host provides energy or food for the parasite. However, this relationship is only beneficial to the parasite. The host is being harmed, whether that is by being deprived of the energy or food consumed by the parasite, or by physical damage to organs, or by chronic inflammation as the host's immune system tries to destroy the parasite.

FIGURE 6.12

Carpenter ant infected with *Ophiocordyceps* spores.

© shunfa Teh/Shutterstock.com

Several parasitic microorganisms actually control the behavior of their hosts by penetrating the blood-brain barrier and directly affecting their nervous systems. For example, *Ophiocordyceps unilateralis* is a parasitic fungus (a eukaryotic cell) that infects carpenter ants and drives them to a higher spot right before fungal spores erupt from their heads to rain down on other ants (figure 6.12). These spores infect the unsuspecting ants and the cycle begins anew. There are thousands of known species of *Ophiocordyceps* and *Cordyceps* (a different genus but close relatives with similar parasitic behavior), and each specializes in parasitizing a different organism. *Cordyceps* was the cause of the zombie outbreak in the Playstation game *The Last of Us* (Sony Computer Entertainment America, 2014). There are several other examples of brain parasites.

*Leucochloridium paradoxum* is a parasitic flatworm that infects snails and changes their eyestalks to look like pulsating caterpillars that attract birds. When a bird eats the infected snail, the flatworm's eggs mature in the bird's digestive system.

*Ampulex compressa* is a parasitic wasp that stings cockroaches and turns them into mindless pack mules. The wasps lead the cockroaches underground and lay their eggs inside the roaches' abdomens. The larvae hatch and eat the cockroaches alive as they mature over the course of a month.

*Sacculina spp.* are a genus of over 100 barnacle species that attach to crabs, castrating the males. These parasites attach themselves to the host shell and siphon energy from it. When *Sacculina* releases its eggs, the mind-controlled crab cares for the eggs as if they were its own.

There are numerous other examples, but you're likely thinking to yourself, "Sure it's one thing to control an ant or a crab, but no way something as complex as the human brain can be controlled like that."

Meet *Toxoplasma gondii*, the causative agent of toxoplasmosis.

FIGURE 6.13.

*Toxoplasma gondii* embedded in muscle tissue

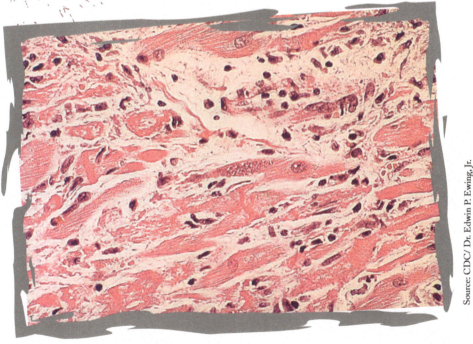

Source: CDC/ Dr. Edwin P. Ewing, Jr.

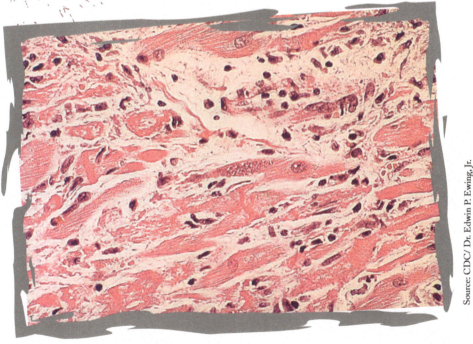

## *Toxoplasma gondii*

*Toxoplasma* is a single-celled protozoan parasite that can affect the behavior of host mammals. Mice and rats infected with *Toxoplasma* have altered behavior, even after the infection is cleared from the mice (in other words, it causes a permanent behavior change). The rodents become foolhardy and reckless… normally, they have an aversion to cat urine, and will avoid areas that smell of cat urine. But rodents infected with *Toxoplasma* show no fear of cat urine and will within 3 weeks of infection even be attracted by it.

The parasite's primary host is the cat that eats the infected rats and mice. It is within cats that the parasite forms eggs, so the hypothesis is that the parasite is altering the rodent behavior to make certain it can reach the primary host, the cat. But it can infect other mammals as well. It's estimated that 30 to 50% of the world's population is chronically infected with the parasite. Luckily for us, it's believed to be an asymptomatic infection (no adverse effects from the infection) except in immunocompromised patients (discussed in Chapter 8).

Pregnant women can be exposed to toxoplasmosis through handling of raw meat (and then not thoroughly washing their hands before eating), consumption of poorly cooked meat, or through exposure to cat feces. If a pregnant woman is infected with *Toxoplasma gondii*, the parasite can infect the fetus across the placenta. In the fetus or newborn, it can cause encephalitis (inflammation of the brain), mental retardation, heart defects, and liver defects. This is why pregnant women are warned by their doctor to avoid changing litter boxes.

A recent study has shown behavior changes in primates infected with *Toxoplasma* parasite (Poirotte, 2016). This study compared behavior of infected and noninfected chimpanzees in response to leopard urine. They found a correlation between chimpanzee infection with *Toxoplasma* and willingness to approach and investigate leopard urine. Scientists have begun looking at humans infected with *Toxoplasma* and behavior studies. Studies have shown infected humans showed prolonged reaction times. Another found a correlation (not causation… we'll go over the distinction in the next chapter) between latent *Toxoplasma* infection and traffic accidents (Flegr, 2002).

New studies show that there is a higher rate of latent *Toxoplasma* infection in schizophrenic patients than in the general population. Current data suggests that most schizophrenia cases are caused by genetic factors or environmental factors (like maternal inflammation during pregnancy), but infectious causes for at least some cases (perhaps

**FIGURE 6.14.**

"Not so innocent now, am I?".

© Top Photo Engineer/Shutterstock.com

up to 20% of cases) seem to be supported by epidemiological data. Most current hypotheses suggest that some combination of genetic background, psychiatric vulnerability, and environmental exposures such as *Toxoplasma* play a determinative role in which patients develop schizophrenia (Hinze-Selch, 2007). Symptoms include severe neural degeneration, including schizophrenia, OCD, and ADHD. Other studies have shown correlation between *Toxoplasma* infection and aggression, suicidal behavior, and sensation-seeking impulses (Cook, 2015). Schizophrenia related to *Toxoplasma* infection has been colloquially dubbed "Crazy Cat Lady Syndrome".

Luckily a zombie outbreak caused by *Toxoplasma gondii* is easy to avoid… don't get a cat, or at least not 20 cats.

MOVIE SPOTLIGHT: *Night of the Creeps* (Dekker, 1986)
**Causative agent—alien brain parasite**
**Also a slasher and alien movie as well as a zombie film**
**All characters share the last name of famous horror directors**
**"I got good news and bad news, girls. The good news is your dates are here."**
**"What's the bad news?"**
**"They're dead."**

# PRIONS

For the record, one of the odds on Vegas favorites to start the zombie apocalypse is the prion. Never heard of it? Well there's no test for it, no vaccine, no cure, and 100% of the cases prove fatal. So maybe we should start paying a little closer attention.

Like viruses, prions are also obligate intracellular parasites. What separates them from every other microorganism we've talked about so far, however, is what they're composed of. A prion is simply a parasitic protein. That's it. Not a cell, no DNA, no RNA, just a single molecule that can parasitize and kill you… and we haven't figured out how to test for it without killing the patient, let alone cure it.

## PRIONS IN THE FOOD CHAIN

With the identification of both **mad cow disease** and **scrapie** (which affects sheep and goats) as **prion diseases** (affecting the same normal neuron protein and causing it to misfold as is seen in Crutzfeldt–Jakob disease), it was realized that the misfolded prion protein had entered the food chain. Because prions are pure proteins that misfold prions can be **denatured** (or unfolded) during sterilization or cooking, and then can refold inside the brain of the creature that ate them, slowly forcing other proteins to misfold and accumulate as a nonfunctional fiber, eventually causing brain degeneration called **spongiform encephalopathy** after the holes it causes in the brain.

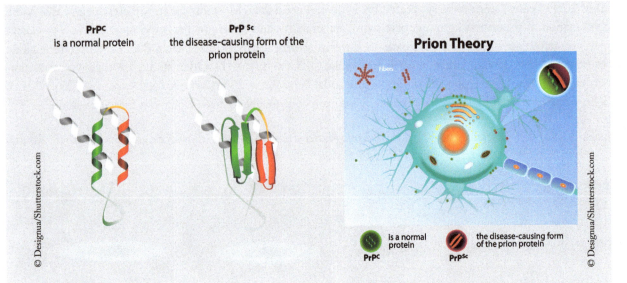

**PrP^C**
is a normal protein

**PrP ^Sc**
the disease-causing form of the prion protein

**Prion Theory**

Fibers

🟢 is a normal protein
**PrP^C**

🔴 the disease-causing form of the prion protein
**PrP^Sc**

© Designua/Shutterstock.com

© Designua/Shutterstock.com

The fact that prions are just proteins (with no genetic information like viruses and bacteria) makes prions uniquely difficult to destroy, and once they are in the food chain, they pose a risk to anyone eating that tissue.

**Chronic Wasting Disease** (CWD) is the version of the prion disease that affects both wild and domestic cervids (deer, elk, and moose). While the United States has been able to limit the spread of Mad Cow Disease in domestic cattle populations through monitoring and careful culling of affected animals, CWD has spread throughout the country in wild cervid populations over the last few decades. It is important for hunters to have samples taken of the animals they intend to eat, and have that meat cleared for human consumption.

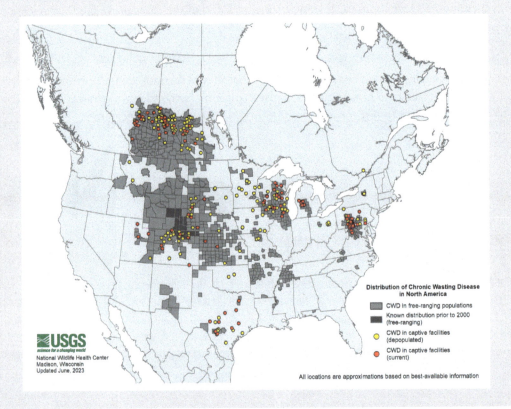

**Distribution of Chronic Wasting Disease in North America**

⬛ CWD in free-ranging populations

⬛ Known distribution prior to 2000 (free-ranging)

🟡 CWD in captive facilities (depopulated)

🔴 CWD in captive facilities (current)

All locations are approximations based on best-available information

**≈USGS**
*science for a changing world*
National Wildlife Health Center
Madison, Wisconsin
Updated June, 2023

Infection from a prion causes a neural degeneration known as spongiform encephalitis. In bovines, we call it mad cow disease (bovine spongiform encephalitis). When it affects humans we call it Creutzfeldt-Jakob disease. Prion-type processes have also been linked to the progression of Alzheimer's disease, Parkinson's, and amyotrophic lateral sclerosis (ALS or Lou Gehrig's disease).

Prions are misshapen forms of a protein found in the neuron cell membrane of healthy mammals. When the infectious, misshapen prion comes into contact with the healthy protein, it causes the healthy to bend and become a prion, which in turn infects other molecules, causing a cascade effect. Eventually so much of this nonfunctioning "junk" protein builds up in a cell that the cell can't function and dies. This causes brain shrinkage and neural deterioration in the victim. The degeneration process to showing symptoms can take years or even decades.

The neural degeneration caused by the parasitic protein leads to changes in gait (walking), hallucinations, lack of coordination (for example, stumbling and falling), muscle twitching, myoclonic jerks or seizures, rapidly developing delirium or dementia. What does that sound like? Zombies. How about the fact that it's transmitted via consumption of infected neural tissue (eating brains – zombies!) or bloodborne (through bites – zombies!!). How about the fact that it was first discovered in flesh-eating cannibals (c'mon, now you're making this up… dude-we totally aren't!!!)

MOVIE SPOTLIGHT: *Zombieland* (Fleischer, 2009)
Causative agent – prion responsible for bovine spongiform encephalopathy
Like 28 Days Later, infected humans, not reanimated corpses
Zombies apparently fooled by zombie disguise (as used by Bill Murray playing himself)
"My mama always told me someday I'd be good at something. Who'd a guessed that something'd be zombie-killing?"

# KURU

FIGURE 6.15.    *Cannibal feast on the Island of Tanna, New Hebrides* Charles E. Gordon Frazer (c. 1885-1889).

In the late 1950s Australian explorers first noted Kuru as another form of spongiform encephalitis, endemic in cannibalistic tribes around Papua New Guinea. These tribes consumed their dead loved ones in order to allow some part of their loved ones to "live on." While stronger male adults were awarded the select parts of the flesh of fallen family members, females and their young were often left with the offal, including brain and spinal cord.  It turns out, it only takes one person in the community to have the random mutation that causes a misshapen prion protein… and due to this custom of the brain being consumed by others… the misshapen prion protein is passed to the eaters, who then develop the illness. As a result, women and children were significantly more likely to become infected and develop Kuru.

Like other forms of spongiform encephalitis, Kuru proved to be incurable and 100% fatal. However, as Australian law enforcement agencies outlawed cannibalism, the infection rates declined then vanished altogether.

… but can they really cause the zombie apocalypse?

# THE WENDIGO

The Wendigo is part of Native American mythology of various Algonquin peoples of the colder northern Atlantic and Great Lakes regions of North America. According to the mythology, one who resorts to cannibalism to stave off starvation is cursed to wander the earth as Wendigo, a creature that constantly craves human flesh. Ojibwe (an Algonquin tribe) scholar Basil Johnston described the Wendigo as such:

"The Wendigo was gaunt to the point of emaciation, its desiccated skin pulled tautly over its bones. With its bones pushing out against its skin, its complexion the ash gray of death, and its eyes pushed back deep into their sockets, the Wendigo looked like a gaunt skeleton recently disinterred from the grave. What lips it had were tattered and bloody [....] Unclean and suffering from suppurations of the flesh, the Wendigo gave off a strange and eerie odor of decay and decomposition, of death and corruption."

Many anthropologists interpret the Wendigo mythos as a warning against resorting to cannibalism, an almost universal taboo among cultures. In the Algonquin cultures, resignation to death or suicide in harsh frigid climates was favored over cannibalism even to survive.

It is possible, however, that the mythos may have had origins rooted in reality as well. The signs and symptoms attributed to the cannibalistic disease Kuru mentioned previously in this chapter can be said to resemble those described by the poor soul possessed by the curse of the Wendigo. It is feasible that individuals that relied on cannibalism to survive may have been afflicted with a neurodegenerative disorder caused by a prion similar to Kuru.

FIGURE 6.16. *The Wendigo* acrylic on canvas by Emily Adele and Ego 2016.

While we currently don't know of a microbial pathogen that is capable of causing symptoms of human zombies, there are many microbes that do have devastating effects that don't quite add up to zombieism. As we've seen, viruses like measles are extremely contagious; viruses like Zika and parasites like malaria can be spread insidiously, by a vector (mosquito) that is almost impossible to completely stop. Behavioral and biochemical changes in the brain can be caused by parasites like *Toxoplasma gondii* or bacteria like *Neisseria meningitidis*. But none of these current pathogens quite add up (yet!) to a cause of zombieism.

Similarly, you've learned in previous chapters that radiation, while deadly at close range, cannot cause the full onslaught of zombie symptoms on its own. Radiation can cause burns that might physically resemble fictional zombies. And radiation exposure at a distance can lead to changes in DNA expression (remember the astronaut study?) and potential mutations in multiple genes, but as we said, mutations in an individual are not contagious.

But… and this is a big but: perhaps the most frightening possibility is that one of these microbes could mutate, or change its genetic information. Maybe radiation exposure leads to mutation of one of these microbes. Maybe viral recombination (when two virus strains shuffle their genes) leads to a virus with a new cellular target (brain cells?!?). Maybe simultaneous infection with a mutated virus that targets brain cells (like a new mutated form of the Zika virus) and a bacterium that affects skin and muscle (Hansen's disease) could lead to zombie-like symptoms.

Remember that protein spikes on the surface of viruses are what allow them to target specific host cells. Remember that proteins on the surface of a parasite or bacteria can help them survive against a host immune system. Can you think of a change in protein expression that might alter the known host cell target, or symptom profile, of an existing microbe? One that might help a virus target a specific set of neurons in the brain, a specific skin layer, or a specific hormone?

MOVIE SPOTLIGHT: *Slither* (Gunn, 2006)
Causative agent—alien parasite
A parasitic worm crash lands on a meteor. Soon a plague of worms overwhelms the town, turning them into zombies and other mutated creatures.
"It's obvious the bastard's got Lyme disease!"
"What?"
"Lyme disease. You touch some deer feces, and then you… eat a sandwich without washin' your hands. You got your Lyme disease!"

# MIND-CONTROL

There are several examples in the animal world of parasites that are able to manipulate the behavior of their hosts.

One example is in the host the banded killifish, which rush together into a tight school when threatened by their predator, the kingfisher. But a parasitic worm that infests the killifish brain (*Crassiphiala bulboglossa*) releases chemicals into its hosts' brains that manipulate their levels of dopamine and serotonin. Scientists have observed a resulting change in behavior, which means the infested fish don't form part of the school when threatened. They also begin to display more erratic movements (zombie-like!) that make them a more obvious target to predators. This appears to allow the parasite to be eaten (along with the killifish) by the kingfisher bird. As it turns out, the parasite must make its way into the kingfisher bird to finish its life cycle, where the eggs are then excreted into water (to be taken in by aquatic snails, where they mature and are then taken in by the killifish).

An example of a parasite that can infect humans (accidentally), but also controls the behavior of one of it's hosts (ants), is *Dicrocoelium dendriticum*. This parasite is primarily a liver fluke in mammals, such as cows and sheep. Its first host is a terrestrial snail, and from there it enters ants. The parasite move into what passes for an ant's "brain," a bundle of nerve cells, and they manipulate those nerves to cause the ant to climb to the top of blades of grass every night, alone, and hang itself there by its jaws. As it repeats this behavior, every night, eventually it is eaten by a cow or sheep, and then makes its way to the liver of the mammal to begin its lifecycle once more. For this particular parasite, no effects have been observed on its mammalian host brains.

With your new knowledge of how the central dogma works (DNA to RNA to protein to trait) and how microbe infection leads to symptoms, can you trace how a mutation in the DNA of a microbe would lead to new symptoms in a human? Perhaps a human in an environment that would also suppress their immune system… like maybe in space (think about the astronaut twin study!)? That sounds like an amazing story/comic/movie!

1. The eukaryotic parasite *Toxoplasma gondii* may affect human behavior by
   a. altering brain chemistry to make people take more risks
   b. damaging the parts of the brain that make decisions
   c. making the human unable to communicate
   d. altering brain chemistry to make people apathetic
   e. altering brain chemistry to make people hungry

2. Sketch and label a general bacterial cell.

3. True or False: Viruses are a complex type of cell. Justify your answer.

4. Which of the following structure determines that an influenza virus particle can only infect respiratory tract cells and blood cells?
   a. protein spikes on surface
   b. the membrane
   c. the enzymes
   d. the capsid
   e. the organelles

5. Describe how prions work to cause brain cell death (spongiform encephalitis).

| WORDSTEMS | |
|---|---|
| strepto- | twist; pliant |
| staphylo- | bunch of grapes |
| -coccus | berry |
| proto- | first; original; primitive |
| -zo- | animal; life |
| endo- | inside |
| intra- | within |

# CHAPTER 6

**WORKSHEET**

## WRITE A BETTER ZOMBIE STORY PART I
## (DEAD, DEADER, DEADEST)

Throughout the semester you're going to attempt to do what Hollywood can't seem to accomplish. Can you come up with a scientifically valid way that zombies could really exist, from the information you've learned? By the end of this class you will hopefully put together enough information to make a feasible story. Legal notice: If you are able to profit off of your story, the authors of this textbook will take a measly 25% of the movie rights, action figures, tee shirt sales, and lunch boxes (yes, that's 25% apiece).

For part I, describe your zombie, and determine whether it is truly dead based on the characteristics of living things. As you learn more throughout the book we can revisit this worksheet and update as necessary.

## WRITE A BETTER ZOMBIE STORY PART II
## (HOW COULD THIS HAPPEN?!)

Write a backstory for how YOUR zombies, in your story could exist. What is the causative agent of the infection? Radiation? Chemicals? Microorganisms? Again, don't pick one with a religious/supernatural mechanism. It's important to be consistent with what you've learned about biology so far. Read back over your story from part I (dead, deader, deadest) and make changes in your zombie's "deadness" as needed.

PLAN YOUR RESPONSE TO PART II HERE

# SECTION 3

## Plan, Prepare, and Persevere

# CHAPTER 7

## THE PANDEMIC

### (OR… "BOB JUST SNEEZED. BETTER PUT HIM DOWN JUST TO BE ON THE SAFE SIDE.")

With contributions from
Shawn G. Gibbs, Ph.D. MBA, CIH
Executive Associate Dean for
Academic Affairs and Professor of
Environmental Health
Indiana University School of Public
Health-Bloomington

After the initial shock of the dead rising, which is then followed by the inevitable scrambling to survive, eventually attention turns to *what* caused this outbreak and *why* this outbreak occurred. Previously in this book we have covered the *what* as you learned about the many different microorganisms that could have caused your zombie outbreak. Now we are turning our attention to the *why* portion to determine how the zombie outbreak is able to spread, and what, if anything, can be done to limit or stop its spread.

## EPIDEMIOLOGY

The Centers for Disease Control and Prevention refers to epidemiology as "… the method used to find the causes of health outcomes and diseases in populations" [CDC, Excellence in Curriculum Innovation through Teaching Epidemiology (EXCITE), 2016]. However, they also refer to the more standard definition that "epidemiology is the study (scientific, systematic, and data-driven) of the distribution (frequency, pattern) and determinants (causes, risk factors) of health-related states and events (not just diseases) in specified populations (neighborhood, school, city, state, country, global)" (CDC, Principles of Epidemiology in Public Health Practice, 2012). In short, these are the professionals who will actively work to identify the microorganisms that have caused the dead to rise as well as all the other factors that contribute to the spread of this infectious disease.

In the case of zombies, the epidemiologist would be involved in what has traditionally been referred to as an outbreak or field investigation. The epidemiologist must work with a multidisciplinary team in order to conduct an appropriate epidemiological study to determine what is going on. The basics of an epidemiological study will involve 1) study design, 2) conducting the study, 3) data analysis, 4) data interpretation, and 5) the communication of study findings (because if you don't tell people what you found out about the zombies then what good is it?!). Essentially, (at least at the early stages while some semblance of society still exists and before the inevitable descent into savage survival), an epidemiologist will work with environmental health experts to collect samples, laboratorians to determine the identity of the microorganism that is causing the disease, and many other experts. They will also conduct survivor interviews and work with the surrounding public to determine how the disease is being spread, which may involve the collection of many eye-witness statements as well as medical histories. Biostatisticians will be brought in to help analyze and interpret the data. Then finally the study results will be communicated. All of this is done to help determine the best and most practical ways to slow or prevent the spread of the disease and prevent it from happening again.

## CORRELATION VS. CAUSATION

To understand the difference between correlation and causation, we need go no further than to paraphrase an example from the seventh season of *The Simpsons* where Lisa explains we must be wary of specious or faulty reasoning (Dietter, 1996).

Homer: "Not a bear in sight. The Bear Patrol must be working like a charm."

Lisa: "That's specious reasoning, Dad."

H: "Thank you, dear."

L: "By your logic I could claim that this rock keeps tigers away."

H: "Oh, how does it work?"

L: "It doesn't work."

H: "Uh-huh."

L: "It's just a stupid rock."

H: "Uh-huh."

L: "But I don't see any tigers around, do you?"

H: "… … … Lisa, I want to buy your rock."

**FIGURE 7.1.**

Tell me you're not humming the theme song in your head.

© 360b/Shutterstock.com

If we were to state that this very book you are reading keeps away zombies because you don't see any zombies around right now, that would be a faulty assumption similar to Homer's.

The mere presence or absence of an item, such as this book, does not necessarily indicate a relationship where that item causes the outcome of instance, such as attracting or repelling a zombie. This can be summarized as follows: Correlation does not imply causation. The word correlation refers in the broadest terms to a dependent relationship between two variables. Whereas, causation means that one variable change causes a change in the other variable. We must understand that two variables can change together, without necessarily having one cause the other, and that to understand this difference requires in-depth study.

An example of this could be dirt and zombies. Think of any zombie movie that you have ever seen, the zombies are most always covered in dirt; if you were operating with specious logic, you might say (like Homer) that it is the dirt that is causing the zombification. However, the dirt didn't *cause* them to become zombies; instead they are covered in dirt because they are zombies and zombies no longer care about personal hygiene or avoiding walking through mud puddles. There is thus a *correlation* between being dirty and zombies, although being covered in dirt does not *cause* one to become a zombie. If one were to observe many uninfected humans through the process of becoming zombies this would be apparent, as you could see them going from being alive and clean, to being zombified and undead and clean, then to undead and dirt covered. This might not be so clear if you are only observing the after-effect, and cannot observe the initial transmission. This is one of the reasons why determining an appropriate study design is so crucial.

# CAUSE OF THE OUTBREAK

Knowing the differences between an outbreak, an epidemic, and a pandemic will allow you to know how widespread an event is, which impacts the likelihood that you will be able to get help from outside of your community. An outbreak is when there is "a sudden rise in the number of cases of a disease" within a defined community or geographical area (Epidemiology, 2016). The terms outbreak and epidemic are synonymous in public health, but usually the term epidemic is used when there is an outbreak of special interest or something. The sentinel or index case is what starts your outbreak and is the first individual to become ill and then pass it along. So, if your friend Bob is bitten by a rat and later becomes a zombie and then Bob bites your sister Stacey, who also becomes a zombie, then Bob is your index case. At this time last year there were zero zombies roaming your town, and this year you have Bob and Stacey currently occupying your backyard—congratulations! You have a zombie outbreak. Now, remember that with diseases that occur more frequently than zombies, such as the seasonal flu, an increase of two cases does not necessarily mean that you have an outbreak. There are often sporadic cases of common illnesses in any given population. You would need to carefully evaluate and compare data to determine if the increase in cases (plus two in our example) more than the expected baseline only then would this constitute an outbreak. This type of data analysis is often done with the help of a biostatistician, a public health professional who is specifically trained to apply statistical techniques to both medical and biological data. Keep in mind that for some diseases, a single case would constitute and outbreak. For example, a single case of Ebola in the U.S. would be an outbreak because our expected baseline of Ebola disease cases is zero. On a side note, if you enjoy biology or medicine along with statistics or mathematics and you want a challenging, well-paying, and in-demand profession then you should consider a career in biostatistics.

Now that you have an outbreak, what might make your outbreak become an epidemic? Most outbreaks of infectious diseases will be self-contained or controlled by local public health interventions and will not move on to become epidemics. However, if your infectious disease outbreak "spreads rapidly to many people," then you now have an epidemic (*Epidemiology*, 2016). If the two zombies, Bob and Stacey, in your backyard escape and quickly become six, and then this spreads beyond your neighborhood to your community, and then on to neighboring towns… well, then your local outbreak just became an epidemic. If we take this one step further, and your epidemic spreads over a larger geographical area or even goes global then you have the makings of a pandemic, and probably the next big blockbuster movie (assuming any part of civilization remains to watch it). How fast your outbreak can grow into an epidemic and then onto a pandemic depends upon a number of factors, including the reproduction rate ($R_0$), the mode of transmission, and the virulence and fatality rate of the microorganism causing the diseases.

# VIRULENCE, INFECTIVE DOSE, INCUBATION TIME, AND FATALITY RATE

The *virulence* of an organism refers to the severity of a pathogen. There are many microorganisms that humans encounter on a daily basis that do not cause disease and as such are not very virulent. There are a number of microorganisms, known as opportunistic pathogens, that are not able to cause disease in otherwise healthy humans, but can cause disease when the body's immune system is suppressed, such as when one has leukemia or HIV. Each species of microorganism has its own infective dose, which is the number of microorganisms required to cause the disease in the host (Leggett, 2012). If the organism that caused the zombie disease had the same infective dose as the Ebola virus, then it would take 1–10 microorganisms to turn someone into a zombie. To give you an idea of how easily this infective dose can be reached, a milliliter of bodily fluid contains thousands to hundreds of millions of microorganisms when someone is contagious with a disease (Public Health Agency of Canada, n.d.). A single droplet from a sneeze can contain many times the infective dose needed for most viruses. On the other hand, a common cause of bacterial gastrointestinal tract illness, *Campylobacter jejuni*, requires ingestion of between 500 and 900 microorganisms in order to become sick (Public Health Agency of Canada, n.d.); however, just to be safe, don't measure out 400 organisms of *Campylobacter jejuni* and drink it. So, the infectious dose for the Ebola virus is much lower than that for the organism that can give you diarrhea or projectile vomiting. If a single bite can transmit our zombification microorganism, it probably has a low infectious dose.

The incubation time (or incubation period) is how long it takes someone to develop the disease once infected with the microorganism. *Campylobacter jejuni* will take 1 to 10 days to cause the symptoms gastrointestinal once you have been infected (Public Health Agency of Canada, n.d.). Whereas a person infected with the Ebola virus will show symptoms anywhere from 1–21 days after infection (Public Health Agency of Canada, n.d.). Some movies have shown an incubation time of days to weeks for zombification; others have shown as few as 10 seconds. Honestly, seconds or minutes is really an unrealistic incubation time for any microorganism; even exposure to the infamous norovirus does not result in symptoms for 24 hours.

The fatality rate (also known as a case fatality rate) for a disease gives an indication of the proportion of the affected population who are likely to die from the disease. The Ebola virus disease has a fatality rate of 50–100%, depending upon which strain of the virus is acquired by the patient (Public Health Agency of Canada, n.d.).

# $R_0$ FACTOR AND THE EPI-CURVE

The reproduction rate, also referred to as the $R_0$, takes into account a number of factors in order to measure the transmission potential of a given disease, and is easiest to think of as the number of secondary infections that result from the initial case (Knowledge, 2016; Ottowa, 2016). So, in our previous example, if Bob only infects Stacey then the zombie disease only has an $R_0=1$; the good news is, unless a disease has an $R_0$ greater than 1, it is very unlikely for an epidemic to occur. However, if on average a zombie like Bob can infect 25 people, then the zombie disease would have an $R_0=25$. Such a very high number indicates a zombie epidemic is highly likely. In order to put this in perspective with known diseases, the Ebola virus disease in developed countries has an $R_0=2$ and measles has $R_0=18$. Measles has a high reproduction number, and so epidemics used to be common; however, we now have vaccines to control measles infections. It is hard to know what $R_0$ our zombification microorganism might have, although our experience has shown us that the $R_0$ of viral diseases is usually higher than the $R_0$ of bacterial diseases. The $R_0$ can really affect our zombie outbreak, determining whether we have a small, contained outbreak or a worldwide pandemic.

If the outbreak moves into an epidemic you will likely see what is referred to as an epidemic curve or epi-curve (Gregg, 2002). An epi-curve is a histogram that graphs the number of cases of the disease on the Y axis against the entire time period of the outbreak on the X axis, starting with day 0 and going though the time period when cases are no longer found (Figure 7.2). This graphic representation is commonly used to demonstrate the length and scale of an outbreak. Both the $R_0$ and the Epi-curve are ways that individuals will use to gauge the scope of an epidemic as well as its potential for growth.

FIGURE 7.2.  The zombie epi-curve (Munz, 2009).

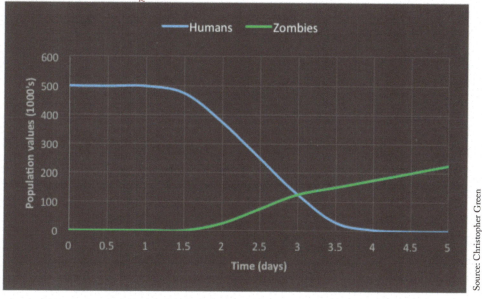

Source: Christopher Green

# UNDERSTANDING TRANSMISSION

FIGURE 7.3. WWII propaganda poster.

There are several routes of disease transmission, including both direct and indirect transmission (Centers for Disease Control and Prevention, n.d.). If the zombie microorganism has to be directly transmitted, then in order to be infected one must have direct contact with the reservoir (person, plant, animal, etc.) or infected individuals for the infectious agent.

According to the CDC, direct contact includes "personal contact, such as touching, **biting**, kissing or sexual intercourse. In these cases the agent enters the body through the skin, mouth, an open cut or sore or sexual organs. Infectious agents may spread by tiny droplets of spray directly into the conjunctiva or the mucus membranes of the eye, nose or mouth during sneezing, coughing, spitting, singing or talking (although usually this type of spread is limited to about within one meter's distance.) This is called droplet spread."

SHE MAY LOOK CLEAN—BUT

PICK-UPS
"GOOD TIME" GIRLS
PROSTITUTES

**SPREAD SYPHILIS AND GONORRHEA**

You can't beat the Axis if you get VD

Source: US National Library of Medicine ca. 1940

**MOVIE SPOTLIGHT:** *Contracted* (England, 2013)
Causative agent—sexually transmitted viral disease
Patient zero originally contracted from necrophilia
Possibly the only movie with an STD as the mode of transmission
"Do not come into contact with anyone until we can determine what it is we're dealing with."

Indirect transmission includes both vehicle-borne (object or material) and vector-borne (insects or other living creature) transmission. In vehicle-borne transmission you may come into contact with a fomite—an inanimate object—that has the infectious organism on it and be exposed to the microorganism. This contact—whether it is visible like touching blood on a baseball bat that was recently used to dispatch a zombie, or invisible like touching a table that infectious particles had settled upon from a zombie passing by—may provide a high enough infectious dose to result in disease symptoms. In vector-borne transmission, you come into contact with an animal or insect that is not ill, but carries the microorganism and then passes it along to you—and then you develop the disease symptoms. In real life, mosquitos are famous for being

FIGURE 7.4.

Anopheles mosquito, vector for malaria.

© Smith1972/Shutterstock.com

disease vectors, passing along the microorganisms that cause malaria, Zika disease, and dengue fever. In most zombie stories, only humans pass on the zombieism disease via direct contact. This is actually a good thing; humans are easier to spot and easier to avoid than small animals or insects. We have a very difficult time controlling vector-borne diseases, because insects like mosquitoes and flies are ubiquitous and almost invisible. A chainsaw or shotgun is pretty much useless against a single mosquito. A zombification microorganism that spreads via mosquitoes would be very frightening, especially once civilization has decayed to the point where we no longer have stores to buy insect repellent or window screens.

## PREVENTING TRANSMISSION

The field of Industrial Hygiene and the Occupational Safety and Health Administration regularly utilizes what is call the Hazard Prevention and Control model; this includes Engineering Controls, Safe Work Practices, Administrative Controls, Personal Protective Equipment (PPE), Systems to Track Hazard Correction, Preventive Maintenance Systems, Emergency Preparation, and Medical Programs (Labor, 2016). All of these systems can be utilized to disrupt the transmission of the microorganism associated with the zombie outbreak, and prevent either the spread of an outbreak or the creation of a new outbreak.

As the name suggests, Engineering Controls utilize facility design or equipment to separate an infectious substance from those you want to protect. So, an anti-zombie wall around your house that keeps the zombies outside of your house would be an Engineering Control, as would a ventilation system that only allows you to breathe filtered air. Hospitals use Engineering Controls when they build special isolation units that filter and sterilize air that a sick person breathes out, for example.

Safe Work Practices are policies and procedures that are followed to keep you safe. If you had a policy that before anyone left the safe zone behind your anti-zombie wall, they had to use a noise generator to draw the zombies to the opposite side of your house from which you planned to exit, this would be an example of a Safe Work Practice. If you had a rotating duty roster, so that each person sent from the safety of your anti-zombie wall to gather food was well-rested, then this would be an example of an Administrative Control; these are other measures utilized to reduce exposure.

Personal Protective Equipment (PPE) are what people tend to think of first when they think of preventing the spread of infectious diseases. Examples would be fluid impermeable clothing, gloves, and face shields to keep zombie fluids off of you if you have to come into contact with a zombie. However, PPE are only as good as your safe work practices in donning (putting on) and doffing (taking off). For example, using a fluid-proof suit while you are killing zombies with a chainsaw is a great precaution, but after all the zombies are dead, if you get just a bit of zombie blood in a cut while you are taking off the suit, all your precaution is for nothing. PPE can also provide a false sense of security if someone is not utilizing them properly. PPE should only be utilized as a last effort if you can't use Engineering Controls or Behavior Controls (like Safe Work Practices and Administrative Controls) to eliminate exposure (Labor, 2016).

Healthcare settings break down the precautions that they use based upon the transmission route of the disease; there are contact, droplet, and airborne precautions (Centers for Disease Control and Prevention, n.d.) (Table 7.1). These precautions are often recognized by the PPE included in them, but they entail Engineering Controls and Behavior Controls also.

**TABLE 7.1**                                                    Precautions for healthcare settings.

|  | Contact Precaution | Droplet Precaution | Airborne Precaution |
|---|---|---|---|
| **Applied to caring for patients with** | Incontinence, wounds, rashes, etc. | Respiratory viruses | Pathogen that is transmitted through the air |
| **Engineering Controls** | Placed in exam room | Placed in exam room with doors closed | Placed in airborne infection isolation room |
| **Safe Work Place Practice** | Perform hand hygiene, clean and disinfect the room | Perform hand hygiene, clean and disinfect the room | Perform hand hygiene, clean and disinfect the room |
| **Administrative Control** | Instruct patient to use separate restroom than others | Instruct patient to wear a facemask | Instruct patient to wear a facemask |
| **PPE** | Gloves | Facemask, gloves, gown, goggles | Respirator, gloves, gown, goggles |

# CHAPTER 7
## QUESTIONS/WORDSTEMS

1. vectors are
   a. insects that transmit pathogens from host to host.
   b. inanimate objects involved in indirect contact transmission of pathogens.
   c. fecal material from infected hosts.
   d. Humans who are contagious via cough, sneeze, etc.
   e. silent carriers of infectious disease.

2. What is the minimum incubation period (incubation time) for any contagious disease?
   a. 10 seconds
   b. 10 minutes
   c. 1 hour
   d. 1 day
   e. 1 week

3. What is the incubation period (incubation time) for Creutzfeldt-Jakob?
   a. 1 hour
   b. 1 day
   c. 1 week
   d. 1 year
   e. 5-10 years

4. Which of the following is not a type of disease transmission?
   a. Airborne
   b. Direct contact
   c. Genetic mutation
   d. Indirect contact
   e. Droplets

| WORDSTEMS | |
|---|---|
| epi- | upon, over, beside |
| pan- | all |
| patho- | illness, disease |
| gen- | generate, begin |
| vir- | poison |

**MOVIE SPOTLIGHT:** *Kingsman: The Secret Service* (Vaughn, 2014)
Causative agent—electronic signal from cell phones
Signal from SIM cards causes everyone within earshot to go into a violent rage by triggering aggression and repressing inhibitors.
"I'm a Catholic whore, currently enjoying congress out of wedlock with my black Jewish boyfriend who works at a military abortion clinic. So, hail Satan, and have a lovely afternoon, madam."

# CHAPTER 7

## WORSHEET

## EPIDEMIOLOGY (GROUP PROJECT)

Adapted with permission from (Green, Clark, Mathis, Barkhurst, & Mansfield, 2015)

## OBJECTIVES

- To explain how transmission of disease occurs throughout a population
- To use data to trace the origin or original carrier of an outbreak (patient zero)

## INTRODUCTION

Diseases can spread through a population by various means. *Contact* transmission is the physical meeting of a source and a new host, often spread through person-to-person transmission. Touching, shaking hands, and kissing are all common examples of this kind of contact. Other examples include direct contact with secretions or body lesions (such as with herpes and boils), intimate or sexual contact (sexually transmitted diseases), and transmission from mother to infant via breast milk (for example, staphylococcal infections). Many disease organisms such as *Salmonella* and *Campylobacter* can even be transmitted through *direct contact* with animals, animal eggs, or other animal products.

*Indirect contact*, a second common mode of disease transmission, refers to transmission from a source to a new host through an intermediary—chiefly inanimate objects (called *fomites*). Thermometers, eating utensils, drinking cups, and bedding are all common fomites. *Pseudomonas* bacteria are commonly transmitted in this manner.

A third mode of transmission, *droplet spread*, occurs when a pathogen is carried on a particle larger than 5 μm. Because this is a large particle, it quickly settles out of the air; therefore, *droplet spread* depends upon the proximity of a new host to a pathogen source, such as a person sick with measles. Sneezes and coughs typically generate the droplets.

## PROCEDURE

In this exercise, an acid (HCl) and a base (NaOH) demonstrate the role of bodily fluids in the spread of the zombie outbreak, and a pH indicator provides the means to identify patient zero.

From *General Biology Laboratory Manual*, 2/E by Christopher F. Green, Krista L. Clark, Karen Mathis, Sue Barkhurst and Jennifer Mansfield. Copyright © 2016 by Kendall Hunt Publishing Company. Reprinted by permission.

## Materials

1. 0.001 M HCl
2. 1 M NaOH
3. 0.04% Phenol Red solution
4. 13 mm test tubes
5. Tube racks
6. Transfer pipets

## Preparation (to be completed by instructor/lab prep staff prior to class)

- Using letters, label the 13 mm test tubes to match the number of students in the class. If you have a small class, you should include "phantom" students in this exercise.
- Put 1 mL of HCl in all but one tube. These are the uninfected.
- Put 1 mL of NaOH in the remaining tube. Note the letter of this tube. This represents the infected individual.

## Spread of the Infection

- Distribute the lettered 13 mm test tubes, one per student.
- Each student takes four 13 mm test tubes. Label each tube 0, 1, 2, and 3.
- Each student put 5 drops of their solutions into the tube labeled "0." SET ASIDE.
- Find a partner to "swap fluids." Record the letter and name of your partner on the data sheet. Using your larger, lettered tubes, each person should draw up some of their own fluid and place 5 drops of their own fluid into their partner's tube. Cap and shake your tubes.
- Before moving on to the next partner, take a sample of your fluids (5 drops) and put this in the 13 mm test tubes labeled "1." SET ASIDE.
- Repeat steps 4 and 5 two more times, but use a new 13 mm test tube each time (the ones labeled "2" and "3," respectively).
- Add one drop of Phenol Red to each student's large 13 mm test tubes and record the result in the last column of your data sheet. Yellow is uninfected and pink/purple is infected.
- Add one drop of Phenol Red to the 13 mm test tubes if tracking the data becomes difficult.
- Record data on the board or MS Excel sheet/copy it into your sheet.
- From the collected data try to determine which tube was infected.

## QUESTIONS

1. Describe how you determined who the original carrier was.

2. How would a real world setting make this task significantly more difficult?

3. Relate the classroom simulation to sexually transmitted diseases. Why is knowing a partner's sexual history important?

4. Define the terms "outbreak," "epidemic," and "pandemic." Provide an example of each that has occurred. Cite your sources in your response. I suggest the CDC for US information and WHO for global information.

| Student | Partner 1 | Partner 2 | Partner 3 | Result (+) or (-) |
|---------|-----------|-----------|-----------|-------------------|
| A |  |  |  |  |
| B |  |  |  |  |
| C |  |  |  |  |
| D |  |  |  |  |
| E |  |  |  |  |
| F |  |  |  |  |
| G |  |  |  |  |
| H |  |  |  |  |
| I |  |  |  |  |
| J |  |  |  |  |
| K |  |  |  |  |
| L |  |  |  |  |
| M |  |  |  |  |
| N |  |  |  |  |
| O |  |  |  |  |
| P |  |  |  |  |
| Q |  |  |  |  |
| R |  |  |  |  |
| S |  |  |  |  |
| T |  |  |  |  |

| Student | Partner 1 | Partner 2 | Partner 3 | Result (+) or (-) |
|---------|-----------|-----------|-----------|-------------------|
| U | | | | |
| V | | | | |
| W | | | | |
| X | | | | |
| Y | | | | |
| Z | | | | |
| AA | | | | |
| BB | | | | |
| CC | | | | |
| DD | | | | |
| EE | | | | |
| FF | | | | |
| GG | | | | |
| HH | | | | |
| II | | | | |
| JJ | | | | |
| KK | | | | |
| LL | | | | |
| MM | | | | |
| NN | | | | |

# CHAPTER 8

## YOUR IMMUNE SYSTEM

(OR... TIME TO RALLY THE TROOPS ON A CELLULAR LEVEL!)

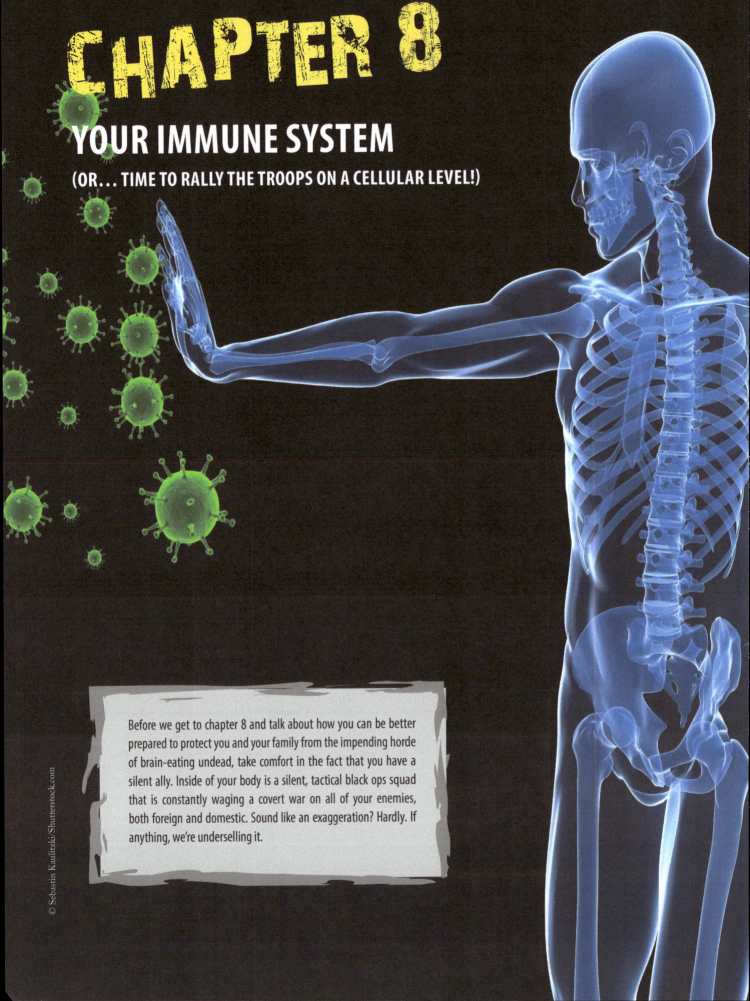

Before we get to chapter 8 and talk about how you can be better prepared to protect you and your family from the impending horde of brain-eating undead, take comfort in the fact that you have a silent ally. Inside of your body is a silent, tactical black ops squad that is constantly waging a covert war on all of your enemies, both foreign and domestic. Sound like an exaggeration? Hardly. If anything, we're underselling it.

So welcome to the front line, grunt. Make no mistake, your enemies are plentiful and will not stop until you've reached room temperature (See what we did there? A reference to chapter 1 homeostasis). Bacteria. Viruses. Fungi. Protozoans. Parasitic worms. They're coming in from every breath you take. Every drink, every bite. Shake a hand? Attacked! Hug a love one? You're hugging a full frontal assault! If all that isn't enough to make you lose sleep, how about our supposed allies, sporadically attacking us from the inside? We have a mutualistic symbiotic relationship with trillions of bacteria in our body. These organisms are called normal microbiota. But our nice, cuddly allies can give way to nasty opportunistic pathogens in the blink of an eye!

## INNATE (NONSPECIFIC) IMMUNITY

For our body, the first line of defense against the causative agents of diseases is called innate immunity; innate meaning the defense was in place at birth. It is also known as your nonspecific defense system, as it defends against each and every foreign invader. These nonspecific defenses includes both physical barriers such as the skin as well as internal cells and chemicals fending off pathogens that get through said barriers.

### Your skin and other protective membranes

Your skin is made of two layers, an outer epidermis of multiple layers of flat cells, and the inner dermis which houses the small organs such as sweat glands, hair follicles, and sebaceous glands.

The epidermis is composed of stratified flat cells called squamous cells. They are stuck together with a substance known as keratin, which gives the skin its waterproof surface. The deepest level of the stratum is composed of stem cells that are constantly dividing. As older cells are pushed up, the keratin prevents oxygen from getting to the cells. As a result the cells die quickly. These tightly packed waterproof cells make it virtually possible for pathogens to penetrate to the dermis *if the epidermis is intact.*

FIGURE 8.1.

Cross section of human skin

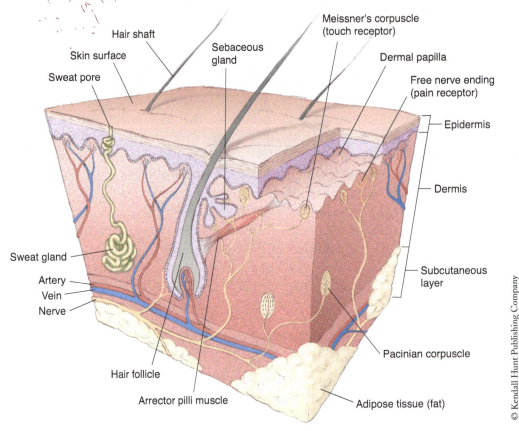

Hair shaft

Skin surface

Sweat pore

Sebaceous gland

Meissner's corpuscle (touch receptor)

Dermal papilla

Free nerve ending (pain receptor)

Epidermis

Dermis

Subcutaneous layer

Sweat gland

Artery

Vein

Nerve

Hair follicle

Arrector pilli muscle

Pacinian corpuscle

Adipose tissue (fat)

© Kendall Hunt Publishing Company

Sweat from sweat glands and sebum (oily secretion) from sebaceous glands (both in the dermis) also help to repel microorganisms. Salt deposited from sweat glands kills microorganisms by pulling water out of cells, upsetting their osmotic balance. Sebum is acidic, effectively inhibiting many microorganisms.

Along with the skin, mucous membranes also protect the internal tissue layers from pathogens. Your digestive, respiratory, urinary, and reproductive systems are lined with cells similar to the keratinized stratified squamous epithelial tissue on the skin. The stratum is thinner than the epithelium and the cells aren't covered in keratin like the skin, but these mucous membranes do afford a significant level of protection.

## DEFENSES IN THE BLOODSTREAM

For those foreign invaders that do sneak past our skin and other protective membranes, you have protective internal components as well. Figure 8.2 shows the cells and cell fragments found in your bloodstream, known collectively as formed elements. These include erythrocytes (red blood cells, which transport oxygen),

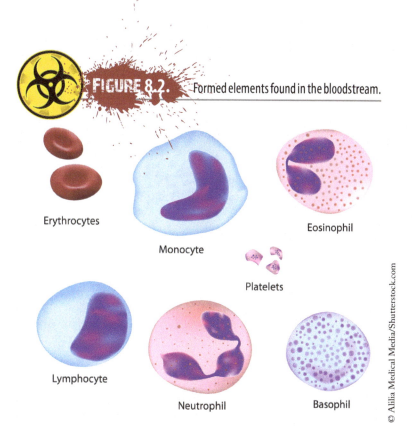

**FIGURE 8.2.** Formed elements found in the bloodstream.

Erythrocytes

Monocyte

Eosinophil

Platelets

Lymphocyte

Neutrophil

Basophil

© Alilia Medical Media/Shutterstock.com

platelets (involved in blood clotting), and leukocytes (white blood cells). There are five types of leukocytes all with different roles of defending against these invaders. Their functions are as follows:

- Basophils cause inflammation, a localized increase in temperature, which helps repair damaged tissue.
- Eosinophils can leave the bloodstream and phagocytize (destroy by engulfing) parasitic worms.
- Neutrophils can also phagocytize invaders and leave the bloodstream as well, or they can kill by producing molecules of hydrogen peroxide and hypochlorite (bleach). Yes, your cells can make bleach.
- Monocytes leave the bloodstream as well and mature into massive (for a cell, of course) macrophages, which eat both foreign invaders and your own dead tissue.
- Finally, lymphocytes are a type of leukocyte with a very particular set of skills. Lymphocytes can be further broken down into three types. Natural killer (NK) cells destroy any of your cells that have been infected with viruses before they replicate and spread. They also destroy tumors located in the body. T lymphocytes and B lymphocytes are involved with your acquired (specific) immunity and will be discussed in the next section.

## CHEMICAL DEFENSES

Your body releases several chemicals to further combat foreign invaders. We will list a few of the more important ones here. Interferons help to protect your healthy cells by interfering with invading virus' ability to attach to them. Complement is a set of chemicals that help trigger inflammation and fever (systemic increase in temperature). Toll-like receptors (TLRs) are located on the surface of phagocytes. They quickly recognize components unique to foreign invaders and kickstart the other components of your immune system. The components located on foreign invaders are called antigens, and are crucial to your specific defenses, as we will cover in the next section. Last but not least, histamine molecules increase blood flow to infected areas to fight infection.

## INFLAMMATION AND FEVER

Though it may be counterintuitive given just how badly you feel when you're down with a fever or with a swollen, red painful hand wracked with inflammation, it turns out both these conditions are actually improving your chances of survival.

First, chemicals such as histamines open up the blood vessels to infected regions. This will in turn bring more leukocytes to the area to phagocytize, or clear out, invaders. An increased number of monocytes will mature into the more effective macrophages and will clear out the infection rapidly. More blood to the area will also bring an increased amount of oxygen and nutrients to the area. These will result in quicker repair of damaged tissue. In addition to speeding up the reaction of your bodies own enzymes, increased temperature has an added benefit of making your body less ideal for certain invaders. For instance, most pathogenic bacteria have an optimal growth temperature of 37°C, your normal body temperature. Every degree increase in temperature lowers the growth rate of these pathogens by exceeding their optimal temperature. Even the pain often associated with inflammation and fever is useful to you. It's your body's way of saying "Don't use the hand, stupid; it's injured."

## MENINGITIS AND ENCEPHALITIS: INFLAMMATION INSIDE THE SKULL

Inflammation can sometimes do more harm than good. Two types of infection response that sometimes occur in zombie origin stories are **meningitis** and **encephalitis**. These are inflammation responses that occur in two different locations, although uncontrolled meningitis can sometimes lead to encephalitis, or they can occur together.

Remember, the suffix -itis means "inflammation". The **meninges** are the protective coverings outside the brain (there are three, the dura mater, arachnoid mater, and pia mater). Infection by pathogens (viruses, bacteria, and even parasites) can cause inflammation of the meninges, and from there the pathogens can move into the brain, which causes encephalitis. Or some pathogens infect the brain through the blood vessels and cause **encephalitis**, inflammation of the brain itself. Inflammation inside the skull is always very dangerous, no matter what the cause, because the skull is stiff, so any inflammation or swelling inside the skull puts pressure on delicate neurons in the brain, which can lead to brain damage.

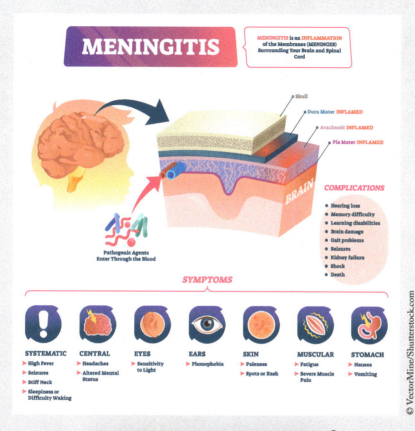

MENINGITIS

MENINGITIS is an INFLAMMATION of the Membranes (MENINGES) Surrounding Your Brain and Spinal Cord

Skull
Dura Mater INFLAMED
Arachnoid INFLAMED
Pia Mater INFLAMED

BRAIN

Pathogenic Agents Enter Through the Blood

COMPLICATIONS
- Hearing loss
- Memory difficulty
- Learning disabilities
- Brain damage
- Gait problems
- Seizures
- Kidney failure
- Shock
- Death

SYMPTOMS

**SYSTEMATIC**
➤ High Fever
➤ Seizures
➤ Stiff Neck
➤ Sleepiness or Difficulty Waking

**CENTRAL**
➤ Headaches
➤ Altered Mental Status

**EYES**
➤ Sensitivity to Light

**EARS**
➤ Phonophobia

**SKIN**
➤ Paleness
➤ Spots or Rash

**MUSCULAR**
➤ Fatigue
➤ Severe Muscle Pain

**STOMACH**
➤ Nausea
➤ Vomiting

© VectorMine/Shutterstock.com

© Stefano Chiacchiarini '74/Shutterstock.com

Meningitis is inflammation of the protective coverings of the brain; encephalitis in inflammation in the brain itself. **Museum of the Megalithic Area of Saint Martin of Corléans, Aosta, Italy – July, 10 2019, skull showing traces of trepanation surgical operation, 3rd millennium, 2nd millennium BC**

Some of the earliest evidence we have of hominids practicing medicine is evidence of **trepanation**, or holes being drilled into the skull. Some theorize that this was done to relieve high pressure in the skull caused by meningitis or encephalitis (or perhaps for other reasons, we just don't know!). We do have evidence that some of these surgeries were successful enough that the patient lived long enough to begin healing by growing new bone where it was drilled away, meaning at least some of the time the practice was survivable.

# ACQUIRED (SPECIFIC) IMMUNITY

While the parts of the body listed above offer an immediate response to foreign invaders, there's another cadre of your defenders that are preparing the counterattack. They are part of your acquired immunity, as you must be exposed to the foreign invader to then acquire defenses specific to that threat. Your specific defense first begins with the lymphocyte known as the T lymphocyte mentioned earlier. One type of T lymphocyte is referred to as a helper T cell. Once a phagocyte (along with the TLRs on its surface) engulfs the foreign invader and breaks it apart, a helper T cell attaches and begins to activate the rest of the immune system, alerting it to the presence of that invader antigen. A helper T cell functions like an emergency dispatcher, letting the rest of your lymphocytes know about the problem. Cytotoxic T cells search for any of your cells that have been infected with a virus and destroy them before the virus replicates.

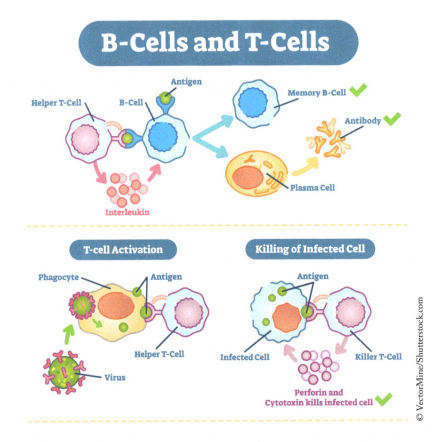

© VectorMine/Shutterstock.com

## ANTIBODIES

Meanwhile, B lymphocytes, your other secret weapon, get the call from the helper T cells. These B lymphocytes are located in the lymph nodes and have the ability to produce antibodies when activated.

An antibody is a special Y-shaped protein that can bind to a specific antigen on the surface of a foreign invader like a lock to a key. You have a million types of B lymphocytes, each type able to make a unique antibody. So, your helper T cell is trying to find and activate the *correct* B lymphocyte to produce antibodies to this particular antigen. Remember, an antigen is a component on the surface of an invader that can be recognized by the immune system and can generate an antibody response.

When the helper T cell activates the correct B lymphocyte, that cell replicates itself to create a clone army, and each member of this army begins to crank out millions and millions of those specific antibodies. The antibodies then travel through the bloodstream to the site of infection, where they bind to the specific antigen on the foreign invader. An antibody can attack a foreign invader by binding to its antigen and causing the invader to agglutinate or clump together. These clumps render the invader inactive and make it an easy target for the phagocytes.

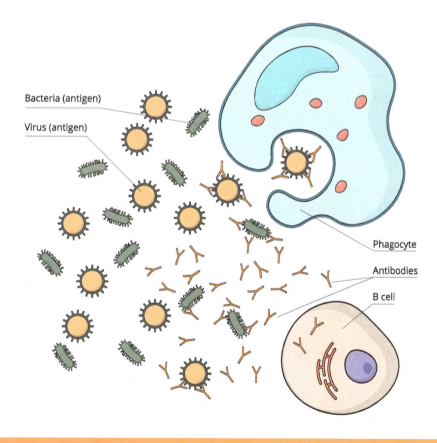

Bacteria (antigen)
Virus (antigen)
Phagocyte
Antibodies
B cell

© Olha Pohrebnyak/Shutterstock.com

## Phagocytosis and B cell

# HIV AND AIDS

The Human Immunodeficiency Virus (HIV) is the causative agent of Acquired Immunodeficiency Syndrome (AIDS). The target (or type of) cell that HIV infects, and eventually kills, are T helper cells. When these T cells drop below a critical level, every part of your specific defenses that rely on signals from the helper T cells becomes inert. Phagocytes aren't called, cytotoxic T cells don't search for infection, B lymphocytes don't know to start producing antibodies. In our metaphor, without the helper T cells as dispatchers, the police and fire departments (or B cells and cytotoxic T cells) don't know there's an emergency, and never leave their stations.

Since HIV effectively stops the specific immune system from functioning, the patient becomes susceptible to infections from microorganisms that would not cause illness in a healthy person. These organisms are called opportunistic pathogens, and we come into contact with them on a daily basis with no adverse effect. AIDS patients can develop infections like pneumocystis pneumonia or severe respiratory tract infections. They also often develop cancers induced by viral infection such as cervical cancer and lymphoma.

Fortunately, medical advances have made HIV/AIDS a chronic disease, rather than a fatal one, in many parts of the world. However, in places like sub-Saharan Africa, where medical care may be harder to come by, the disease is still ravaging the population and causing untold deaths.

## IMMUNE SYSTEM MEMORY

The most remarkable characteristic about your specific defense, however, is its ability to adapt. Your specific immune system has a memory, and the same tactics rarely work on it twice.

After the T and B lymphocytes have done their job eradicating the threat from the foreign invader, a small portion of them become memory T cells and memory B cells. These memory cells can survive for as short a time as one year up to as long as many decades, depending upon the antigens to which they are exposed. The memory cells circulate in the bloodstream and tissue, ready to be reactivated immediately during reexposure to the same antigen.

When they are reactivated by exposure to that same surface antigen, memory cells can clone themselves very quickly and produce an army of responding T and B lymphocytes, producing higher levels of antibodies much faster. They can neutralize an invader in as little as 48 hours—usually so quickly that you never feel any symptoms at all. As a result, there are many illnesses you usually only get once (like chickenpox, caused by the Varicella Zoster virus).

Unfortunately, however, some viruses and some bacteria mutate rapidly, so that by the time you get exposed again, the antigens on the surface of the invader have changed enough so that your memory cells do not recognize it. This happens with cold viruses. There are hundreds of different viruses that cause colds, and they mutate very rapidly.

So, sometimes, you get a cold, and then give it to your roommate, parents, or kids, and then three weeks later you catch it again… because it has changed enough that your antibodies don't stick to it any more, and your memory cells don't recognize it.

## TRIGGERS FOR AUTOIMMUNE DISEASE

**Autoimmune disease** ("auto" = "self") occurs when one's immune system begins to damage one's own cells. **Type 1 diabetes** is a type of autoimmune disease (where one's immune system begins to attack and destroy insulin-producing cells in the pancreas); so is multiple sclerosis (where one's immune system attacks the protective coating around neurons in the brain and spinal cord). Genetics contributes to some autoimmune diseases. Infections can also trigger autoimmune disease, when the proteins on the surface of a virus or bacteria look enough like the proteins on self-cells to trigger a cross reaction.

Viruses known to potentially cause autoimmune diseases include the mumps virus, measles virus, SARS-CoV-2, and Epstein-Barr virus (which causes an illness commonly called "mono," "mononucleosis," or "glandular fever"). Vaccination against mumps and measles decreased the rates of autoimmune diseases due to those illnesses; scientists are now working on a vaccine for Epstein-Barr virus to help further protect people against the autoimmune diseases it triggers.

### AUTOIMMUNE DISEASES

Multiple sclerosis

Hashimoto's thyroiditis

Asthma

Systemic lupus erythematosus

Celiac disease

Rheumatoid arthritis

Eczema and psoriasis

© Designua/Shutterstock.com

### MULTIPLE SCLEROSIS

*Healthy*

*Nerve affected by MS*

Damaged myelin

Node of Ranvier

Exposed fiber

Schwann cells

Nerve fiber

© Designua/Shutterstock.com

**MOVIE SPOTLIGHT:** *Train to Busan* (Yeon, 2016)

Causative agent—unspecified infection; likely stemming from an industrial plant leak.

A group of terrified passengers try to outrun the spreading zombie apocalypse on a train to Busan, a supposed safe zone.

Early in the film, an infected deer is seemingly fatally hit by a truck. It then gets back up and is shown to be infected. So the causative agent might be contaminated venison (commonly consumed in South Korea). It is possible the cause is the prion responsible for *Chronic Wasting Disease.*

♪"Aloha 'oe, aloha 'oe, until we meet again..."♪

1. Which of the following is not part of your acquired (specific) immunity?
   a. Memory B Cells
   b. Helper T Cells
   c. Memory T Cells
   d. NK Cells
   e. Cytotoxic T Cells

2. Which of the following is not part of your innate (nonspecific) immunity?
   a. Epidermis
   b. Fever
   c. Inflammation
   d. NK Cells
   e. B cells

3. Cells that destroy your own cells infected with a virus before they can rupture are called
   a. Memory T cells.
   b. Memory B cells.
   c. Helper T cells.
   d. Cytotoxic B cells.
   e. Cytotoxic T cells.

4. Which of the following control immune response?
   a. Memory B Cells
   b. Helper T Cells
   c. Memory T Cells
   d. NK Cells
   e. Cytotoxic T Cells

5. Which is not a characteristic of the skin that makes it inhospitable for microorganisms
   a. Layers of dead skin cells
   b. Waterproof keratin wax
   c. White blood cells interspersed throughout the skin cells
   d. Salt secreted by sweat glands
   e. Acidic oil secreted by sebaceous glands

6. The concept that vaccines are mainly to protect immunocompromised members of a community is known as…
   a. Morbidity.
   b. Mortality.
   c. Herd immunity.
   d. Epidemiology.
   e. Opportunistic pathogenicity.

7. Why are vaccines ineffective if you're already sick?

8. Why are antibiotics ineffective against viral infections?

9. Explain how antibiotic misuse/overuse leads to antibiotic resistance.

| WORDSTEMS | |
|---|---|
| -cyte | hollow place; cell |
| -dermis | skin |
| erythro- | red |
| leuko- | white |
| phago- | to eat |
| macro- | large |
| plat- | flat |
| lymph- | water |

# CHAPTER 8

## WRITE A BETTER ZOMBIE STORY PART III
## (THE SPREADING APOCALYPSE)

Continue your story by describing the mechanism for transmission. How is it spread and what is the rate of the spread? Again, reread parts I and II and make changes as needed for consistency.

_____

_____

_____

_____

_____

_____

_____

_____

_____

_____

_____

# CHAPTER 9

## TREATMENT

(OR..." IF YOUR URGE TO TEAR FLESH FROM BONE AND SINEW CONTINUES FOR MORE THAN FOUR HOURS CONSULT YOUR PHYSICIAN")

"So it's an absolute lie that has killed thousands of kids. Because the mothers who heard that lie, many of them didn't have their kids take either pertussis or measles vaccine, and their children are dead today. And so the people who go and engage in those anti-vaccine efforts -- you know, they, they kill children. It's a very sad thing, because these vaccines are important."

—Bill Gates, entrepreneur and philanthropist.

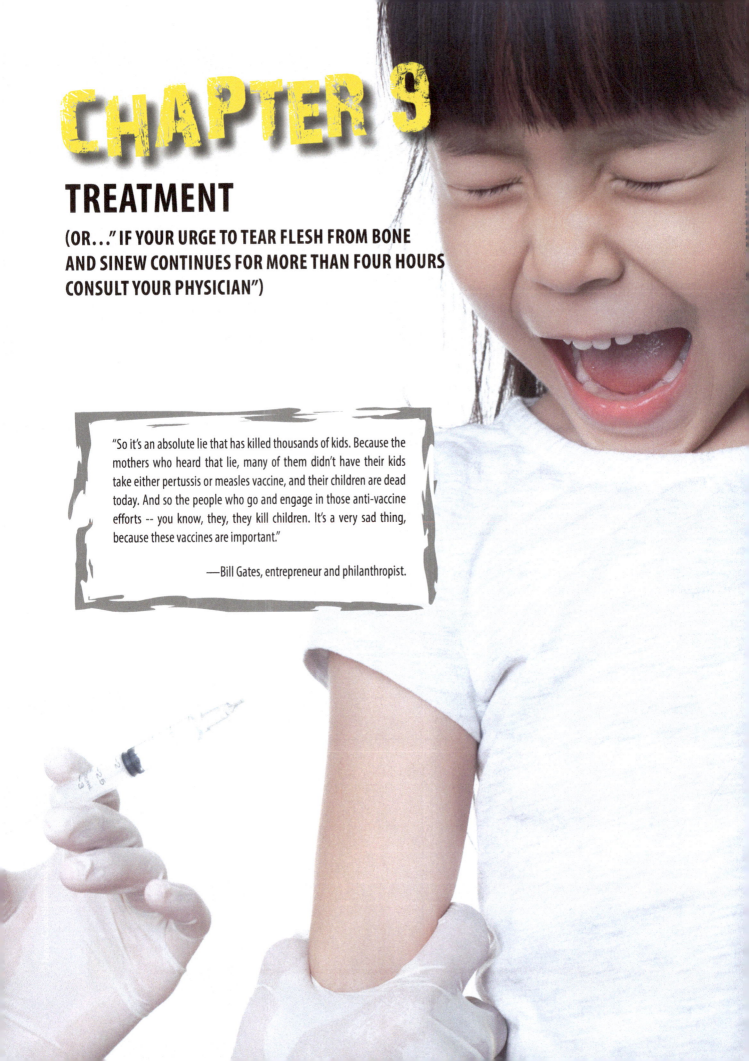

# CONTROL OF PATHOGENS

One of the best ways to avoid illness, whether that is from the flu or from the zombie outbreak, is to avoid exposure to the pathogen at all.

We have a few ways of reducing germs in our environments, which we use to try and control our exposure to disease causing pathogens. But remember, viruses, bacteria, and even eukaryotic pathogens are **ubiquitous**, are present everywhere around us and even, in some cases, inside us.

In our homes, schools, and professional environments, we want to **disinfect** when we can. Disinfection means to reduce the numbers of germs around us and on surfaces. We can do this by cleaning with soap or with disinfecting chemicals. In hospital and healthcare environments, there are times when we want to **sterilize** areas or equipment. Sterilization means to completely eliminate germs on a surface or area.

## WATER AND SEWAGE

Safe water to drink is a luxury that many of us don't think about until a natural disaster or contamination event risks our water supply. Drinking water treatment plants use chlorine kill germs and filtration to remove contaminants before it reaches homes; sewage treatment plants then use multiple stages of filtering, aeration, beneficial bacteria that eat sewage, and final filtration and disinfection to decontaminate sewage so that water can be returned to lakes and rivers without passing on dangerous pathogens.

# SOAPS, DISINFECTANTS, AND AUTOCLAVES

It is often said that bacteria are ubiquitous, which means they are everywhere; in fact, many "good" bacteria are inside our bodies and on our skin, and can even help with essential life functions, like helping us digest food and make vitamins. We often call these bacteria "normal flora," and talk about the human "**microbiome**"; a collection, an ecosystem, of microbes that live on us and in us and help us survive. We are just now scratching the surface of how important these microbiome bacteria (and perhaps even viruses!) are for our health.

But because of pathogens, we have had to find ways to control bacteria and viruses (and now prions) in our environments and especially our hospitals.

Soap is the first line against germs; soap is **amphipathic** (it has a hydrophilic end and a hydrophobic end) and it lifts and separates bacteria and viruses away from our skin, allowing water to wash them away. It is important to scrub with soap for about twenty seconds, to give the soap molecules time to lift the germs away from the skin!

We can also use chemical disinfectants, that often use basic pH or chemicals that break apart bacterial and viral proteins or membranes. **Chlorine** works to disinfect water by pulling electrons away from the proteins and membranes of germs, destroying them. Chemical **detergents** also work this way, pulling apart bacterial cell membranes and virus envelopes.

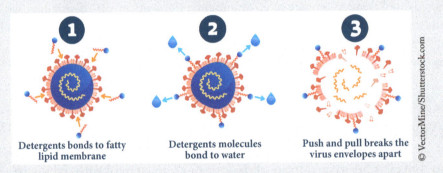

| **Detergents bonds to fatty lipid membrane** | **Detergents molecules bond to water** | **Push and pull breaks the virus envelopes apart** |

© VectorMine/Shutterstock.com

Alcohol-based hand "sanitizer" works to disrupt viral envelopes using alcohol, so works well for influenza viruses and covid viruses, but doesn't always work as well for bacteria or naked viruses. Because it doesn't really get rid of all germs, a more accurate description for these alcohol-based chemicals would be a "hand disinfectant."

Hospitals have special, stronger chemicals to use that aren't as safe for casual exposure. Health professions also use high temperature, steam, and pressure, in special machines called **autoclaves**, to destroy bacteria and viruses on equipment and materials.

Prions are a special case, which requires special sequences of chemicals and steam and pressure, to destroy them. Just cooking meat that contains prions does not destroy the prions, which is how eating cooked meat is still able to transfer the prions and cause symptoms many years later. Remember, prions are not living cells like bacteria, and they contain no genetic information the way viruses and bacteria do. They are just proteins, which can fold incorrectly and cause other proteins to fold incorrectly. Proteins are hard to completely destroy, which is why there are special protocols for destroying them.

There have been cases in the past of sterilization protocols NOT being followed completely, and brain surgery instruments then being used that could transfer the prions from one patient to the next.

In the event of an actual zombie apocalypse, or really any natural or unnatural disaster, it is critically important to have clean drinkable water. Illnesses that spread through water include cholera (a bacterial illness that produces a toxin that causes severe vomiting and diarrhea, and can be fatal), dysentery, typhoid fever, hepatitis, and even polio.

In an emergency, it is possible to buy filtration devices that can filter pathogens out of otherwise untreated water, and it is also possible to use a dilution (very diluted!) of bleach or water treatment tablets to treat water and make it safe enough to drink or cook with.

Control of sewage in an emergency is another important factor; temporary toilets often use bleach to help destroy viruses or bacteria to limit exposures.

## Personal Protective Equipment

We can also control our exposure to germs by protecting the areas of our body where germs get inside us. This can include coverings over our mouths and nose, barriers to protect our eyes, and an extra layer of protection over our skin. This is why scientists working in labs wear lab coats, why medical personnel wear medical gowns or lab coats, and why surgeons wear masks and gloves. PPE is especially effective for pathogens spread by contact, large droplets, or aerosol (small) droplets.

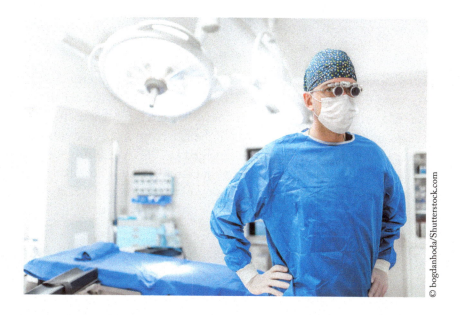
© bogdanhoda/Shutterstock.com

Interestingly, all PPE works better if it is worn by the sick or contagious person, so that they aren't releasing germs into the environment. This is actually why surgeons wear masks and gloves, even in wartime situations; they are trying to prevent adding germs into the patient. But asking a zombie to mask up may be a bit of a stretch. In that case, wearing the PPE to protect yourself can help, by reducing the number of germs entering your body, called an **infectious dose,** or **dose of inoculum.** It turns out, some pathogens need a high dose of inoculum to cause symptoms (for Streptococcus pyogenes to cause strep throat, for example, it's more than one thousand bacterial cells, whereas of measles virus it is a single virus). Wearing PPE reduces the number of pathogens that reach your body, and so can prevent infection. There is some evidence that the severity of covid symptoms could be related to the initial exposure dose (Guallar et al.).

# VACCINES

One of the best ways to prevent infection is to use the ability of the immune system to remember pathogens to our advantage. Vaccines are weakened or inactivated forms of pathogens that are purposefully administered to give specific immunity. They function by making use of the memory capability of the immune system. When a patient receives a vaccine, the antigens on its surface trigger their specific defenses and induce the same immune response one would get if he or she was infected with the pathogen.

This exposure to the antigens creates memory B cells and memory T cells. That means if the patient encounters the pathogen naturally while those memory cells are still viable, the memory cells will quickly activate, creating billions of antibodies and stopping the pathogen before the patient even feels any symptoms. The flu vaccine, measles, mumps, and rubella (MMR) vaccine, and polio vaccine are examples of vaccines against viruses. The whooping cough (pertussis) vaccine, the meningitis vaccine, and pneumococcal vaccine are examples of vaccines against bacteria.

It can take decades to create a vaccine that works (and you have to get pretty lucky). Then it can take decades more to test it and make sure it is safe. As far as Hollywood goes, too many zombie stories have a hero that "manufactures" a vaccine in a matter of days or weeks, and this becomes the miracle that saves the world. That would be great. But as we saw in 2014 with the Ebola outbreak... scientists have been working on developing an Ebola vaccine since the virus first emerged in 1968. We're still not there, although some vaccines that had been developed were fast-tracked into human trials. So... how likely is it that our hero can manufacture a vaccine for our zombie outbreak in a matter of days... or weeks... or even months?

## HERD (COMMUNITY) IMMUNITY

It is a common misconception that babies get vaccination shots at birth or 1 month old, and are then protected fully. In actuality, babies get their first pertussis (whooping cough) vaccine for example, at 2 months; but full immunity for pertussis doesn't happen until they have three shots (finished at about a year old for most kids). Many vaccines aren't given to babies until they are 1 year of age.

FIGURE 9.1.

No vaccinations.

© Emily Adele

That means that newborns and babies are much more vulnerable to infection by a wide range of illnesses than older children or healthier adults. Until they get their full regimen of vaccines we say these babies are immunocompromised, meaning they are far more susceptible to illness. This is a problem because many common diseases that cause mild symptoms in adults cause severe symptoms in young children and babies. Measles, for example, can cause deafness, meningitis, brain damage, and death in babies. Whooping cough, which can cause mild symptoms in adults, causes thickening of mucus leading to terrible coughing spasms that can cause lack of oxygen, brain damage, and death in babies.

Babies aren't the only ones who may be immunocompromised. Organ transplant recipients, for instance, must take medicines to prevent rejection of their new organ. These medicines suppress their immune system, leaving them vulnerable to infection. Cancer patients or HIV-positive individuals also have weakened immune systems and are also immunocompromised. Finally, the very elderly often also are immunosuppressed.

How can we protect these vulnerable people? By something called herd immunity.

It turns out that if around 85% of people in a population are immune (due to having already caught the disease and recovered, or being vaccinated), then more vulnerable immunocompromised people in the population (like newborns, babies, and cancer patients) are less likely to come in contact with the pathogen. When immunization drops below 85% the chances of epidemics in the population increase dramatically. The "magic number" or percentage needed is dependent upon how contagious a given pathogen is. For example, measles is one of the most contagious viral diseases known, and to protect infants under 1 year old and other immunocompromised patients, 92% of the population must be immune or vaccinated.

## ANTI-VAXXERS, THE ZOMBIE'S FIRST COURSE

Then there is that special breed of future Darwin Award winner that decides "Nah, I don't buy it. We don't need vaccines."

On February 28, 1998, the prestigious British medical journal *The Lancet* published an article by Wakefield et al. titled "Ileal-lymphoid-nodular hyperplasia, non-specific colitis, and pervasive disorder in children" (Wakefield, 1998). In it, the authors linked administering the MMR vaccine with increased likelihood of autism development in children. Autism is a developmental disorder that affects patients' communication, interaction, and overall behavior. As a parent it's frightening to think that you could give your child such a developmental

**FIGURE 9.2.** Insufficient vaccinations in the population to achieve herd immunity.

© Emily Adele

**FIGURE 9.3.** Herd immunity.

© Emily Adele

disorder simply by trying to protect him or her with vaccines.

The only problem? The authors made it up. Did they incorrectly interpret their results? Perhaps overreach with their conclusions? Perhaps committed some statistical error?

Nope. Made it the #$&#%!!! up.

Remember what we said about people with agendas presenting information to you. The now former Dr. Wakefield purposely mislead the general public in order to profit from autism tests. By 2010 *The Lancet* retracted the article and Mr. Wakefield lost his medical license.

Unfortunately, the damage was done. Due to his false claim, vaccination rates in the United States and Britain dropped. The more medical experts fought to prove there was no link, the stronger peoples' belief in the link was. Celebrities were suddenly weighing in, also railing against the evils of vaccination.

As a result, there was an increase in serious illness and death rates with respect to measles, mumps, and congenital rubella, as well as the reemergence of other diseases long controlled in the population. Let that sink in. Children died due to this man's lie. Children continue to die because of this man's lie.

Anti-vaccination is nothing new, however. The first vaccine for smallpox was given in 1796 when Edward Jenner inoculated an 8-year-old boy (??!!) with cowpox, then exposed him to smallpox (smallpox variolation was routinely practiced at the time). When the kid didn't die and didn't get smallpox, and Jenner avoided an involuntary manslaughter charge, the first vaccine was born. Not long after, the anti-vaccination movement was born.

**FIGURE 9.4.** Any resemblance to actual persons, living or dead, is purely coincidental.

© Emily Adele

**FIGURE 9.5.**

A 1930 vaccination cartoon depicting anti-vaxxers.

Source: Library of Congress

So while we can't blame the start of the anti-vaxxer movement on Wakefield the deadly consequences of his actions are a stark reminder of the duty of medical and scientific professionals to accurately report their findings without spin or bias.

There are some possible risks of vaccines to a very small, select group of people. A few people might have immunodeficiency, where their immune systems will not produce memory cells from a vaccine, and might instead suffer side effects. One common worry of anti-vaxxers is a side effect where the patient's T cells begin to attack the patient's nerve cells, causing muscle weakness and sometimes temporary paralysis. This is called Guillain-Barre syndrome (GBS).

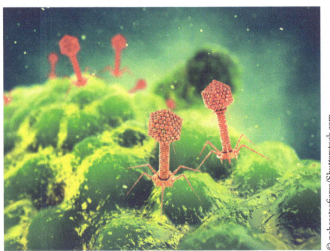

© nobeastsofierce/Shutterstock.com

Multiple studies have shown no risk of increased GBS from flu shots. Interestingly, two studies did find a slight increase in risk of GBS from the H1N1 vaccine (1 new case per 1 million vaccinations), but there was a much higher risk from actual infection with the pathogen H1N1 (17 cases per 1 million vaccinations) than from the vaccine. So again, any risk of GBS is *much* higher in those that are not vaccinated, than in those that receive the vaccine.

# ANTIBIOTICS

Finally we come to your secret weapon; your submicroscopic smart bomb; antibiotics. Antibiotics are chemical substances produced by microorganisms which have the capacity to inhibit prokaryotes in dilute solution.

Did you catch all that? There are a few crucial pieces of information in there, but we can boil it down to a simplified idea. Antibiotics affect bacteria while leaving your cells undamaged. Regular disinfectants like hydrogen peroxide, ethyl alcohol, or calcium hyperchlorite (bleach) are equal opportunity weapons. They'll destroy any cell they can get their hands on. Antibiotics, however, will target prokaryotes *yet do no damage to eukaryotes at that same effective dose.*

Pretty cool, right?

Let's break the definition down further:

"… produced by microorganisms…"

Antibiotics are chemicals or proteins made by other cells (usually eukaryotic fungi) that have to compete with bacteria for food, and so evolved these chemicals to fight off their competitors. Penicillin, for instance, was the first antibiotic to be discovered by modern medicine, by a scientist named Alexander Fleming in 1928. He noticed a green mold contaminating a petri dish where he was growing colonies of *Staphylococcus aureus*, the causative agent of diseases such as staph infections. He noticed no bacteria would grow anywhere near the mold. The green mold was *Penicillium chrysogenum*, and identified the chemical it was secreting to damage bacteria as penicillin. It turns out there is evidence that some ancient civilizations were already working with antibiotics, although they probably didn't understand why they worked. A medicinal beer was used by ancient Nubians that has been shown to contain tetracycline, one of our modern antibiotics, and ancient Egyptians and Jordanians recorded the use of specific recipes of beer medicinally, also (Nelson, 2010).

"…in dilute solution."

Again, this is quite different from a regular disinfectant. Bleach works wonders destroying bacteria but you wouldn't want to inject yourself with it. Any concentration of disinfectant weak enough to safely inject would be far too weak to kill any microorganisms. Antibiotics, however, *work in dilute solution*, meaning they are effective at doses that are diluted to be safe for human treatment.

And finally the most important part,

"…the capacity to inhibit prokaryotes…"

How do antibiotics only target prokaryotes? They work specifically on the parts of bacteria that eukaryotes like us lack. For example, penicillin prevents bacteria from forming their cell walls, which causes them to burst from osmotic pressure. Our cells don't have cell walls, and so are unaffected by penicillin. Other antibiotics destroy the smaller ribosomes bacteria need for protein synthesis. Our eukaryotic ribosomes are larger and are unaffected.

See? Smart bomb.

Well… except that bacteria are living cells, and that means they will have random mutations in their DNA. Every once in a while one of these random mutations changes the antibiotic target slightly. Maybe not enough to render the target unable to do its own job in the bacteria, but just enough to make the antibiotic not recognize it. When this happens we call it antibiotic resistance… and it's a problem. Worse, these mutations can then be passed between bacteria to spread the antibiotic resistance to other bacterial populations.

## ANTIBIOTIC RESISTANCE

So, let's say we have a patient with a urinary tract infection.

We begin treating this patient with a specific antibiotic, penicillin. Penicillin binds to and freezes an enzyme that helps bacteria build their cell wall. Without their cell wall, they die.

One of these bacteria has a random DNA mutation which changes the enzyme that helps build the cell wall. This little guy has no problem building his cell wall, even if penicillin is present.

Because he is able to survive so well in the presence of penicillin, he can grow and reproduce himself rapidly, while all the old bacteria die off.

> FIGURE 9.7.

Realistic representation of a urinary tract infection

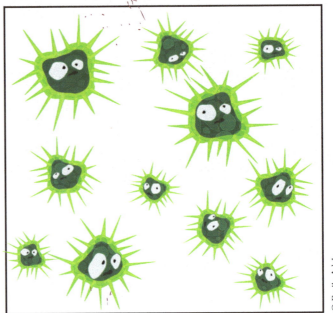

© Emily Adele

> FIGURE 9.8.

Spot the mutated bacterium.

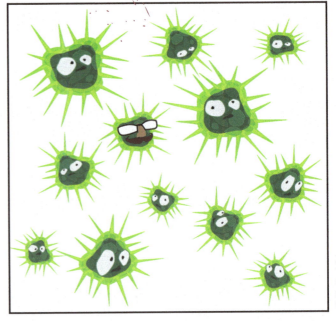

© Emily Adele

For a few days, the patient feels better. But the use of the antibiotic is artificially *selecting* for the bacteria with the mutation that makes it resistant to the antibiotic. After a few days, our little resistant bacterium has reproduced, and now the patient has a UTI caused by all his descendants. They are all resistant to penicillin, and now a doctor must try a different antibiotic.

There is a real danger when we use antibiotics indiscriminately. For instance, 70% of otitis media (inner ear) infections will clear up by themselves after 3 days. However, some pediatricians will still prescribe antibiotics immediately, usually due to the insistence of parents. Sometimes antibiotics get prescribed for viral infections, even though antibiotics only affect prokaryotes and are useless against viral infections.

On top of misuse and overuse of antibiotics in the health care setting, antibiotics in agriculture are also a concern. Over 70% of the antibiotics in the United States are used in livestock, including beef, pork, chicken, and even fish. These antibiotics make their way into the food chain, the water, the soil, even the air. In all of these locations, the antibiotics are then *selecting* for antibiotic resistant bacteria. Antibiotic use in agriculture directly leads to the rise of antibiotic resistant bacterial strains, like MRSA (methicillin resistant *Staphylococcus aureus*), CRE (carbapenem resistant *Enterobacter*), VRE (vancomycin resistant Enterococci), CRKP (carbapenem resistant *Klebsiella pneumonia*), and MDR-TB (multidrug resistant tuberculosis).

**FIGURE 9.9.**

Our mutant survives the antiobiotic attack.

© Emily Adele

**FIGURE 9.10.**

Then comes the resurgence.

© Emily Adele

## OPPORTUNISTIC PATHOGENS

There are other worries about antibiotic use. As mentioned earlier, your body includes normal microbiota in locations like your skin, gastrointestinal tract, and the walls of the urethra and vagina. There are trillions of these bacteria in every human. Typically, we're glad they're there. They prevent colonization of pathogens by taking up all the space and nutrients. They aid in digestion by helping to reclaim water from waste in the large intestines. They can also make essential vitamins K and $B_{12}$ for us.

However, the antibiotics you may be taking to fight infection can also damage your normal microbiota. Remember, antibiotics attack prokaryotic cells, whether they're the *Streptococcus pyogenes* eating the back of your pharynx and giving you strep throat, or the *Escherichia coli* in your intestine aiding in digestion. So when antibiotics wipe out the normal microbiota, there are all kinds of nasty consequences. The most serious is that since they aren't where they should be, microorganisms that typically can't harm you can become dangerous opportunistic pathogens. Patients often experience diarrhea, yeast infections, and UTIs after a dose of antibiotics. More serious opportunistic pathogens can even result in life-threatening illnesses. *Clostridium difficle* is one such opportunistic pathogen. The normal microbiota can usually keep it from taking up residence in your GI tract, but in their absence it is the causative agent of the potentially life-threatening pseudomembranous colitis, otherwise known as C-diff.

# VIRUS TREATMENTS

## WHY ARE VIRAL INFECTIONS HARD TO TREAT?

Antivirals

Treatment of viral infections is tricky, because viruses get inside host cells to reproduce, so anything we use to target viruses potentially has the ability to disrupt or even kill the human cells they are hiding in. This means almost all antiviral medicines we have cause side effects; side effects which are sometimes worse than the infection itself. (Note this is very different from antibiotics, which specifically target bacterial cell structures and mostly leave our cells completely alone.)

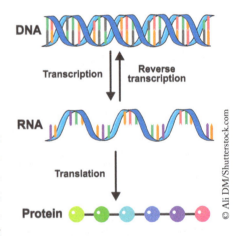

(Human Immunodeficiency Virus) is a retrovirus.

One really successful class of antivirals we have are reverse-transcriptase inhibitors. Some viruses contain RNA as their genetic information, and when they invade host cells they use a special virus enzyme called **reverse transcriptase** to copy their RNA backwards into DNA to invade host cells.

Because of this backwards step, these viruses are called retroviruses; HIV (Human Immunodeficiency Virus) is a retrovirus.

Because this enzyme is only used by viruses in our cells, we can specifically interfere with the enzyme reverse transcriptase without harming the host cells.

Similarly, antiviral treatments used for other viral infections are directed specifically at enzymes that play a role usually unique to the viral infection, in order to target the viruses and not the host cells; it's important not to hurt the patient's cells!

Because viruses reproduce by the hundreds or thousands in infected cells, for antivirus medicines to be effective, they need to be taken <u>very soon after infection</u>. Usually within a matter of days. It is common for there to be a "rebound" after antiviral treatment; if not all the virus particles are able to be eliminated by the immune system during the treatment time, sometimes a single virus particle is enough to restart the infection process after the medicine is stopped.

Survivor Plasma; treating viruses by using antibodies

When a virus is brand new on the scene, and we don't know much about it, we don't' know which enzymes to target to best help patients. One thing we can do is use the blood plasma from patients that have survived the same infection. Remember, when a patient's T and B lymphocytes fight off a pathogen they create memory T and B cells of their own. We say that this is **active immunity**. If that immunity was acquired by the patient actually becoming ill with the disease, we call it **naturally acquired active immunity**. If acquired by a patient's immune system being trained by a vaccine, it's called **artificially acquired active immunity**.

It is actually possible to directly treat a patient with antibodies specific to that pathogen that were made in another person (a survivor of that same pathogen), another animal (like a rabbit or a horse), or even a genetically engineered plant. As the patient doesn't directly (or actively) make his or her own antibodies, we call this **artificially acquired passive immunity**.

So when the donor originally got sick from the pathogen, antibodies specific to that virus are produced by B cells and released by the millions into the blood. (Remember, this primary response can take days or weeks!) Survivors can donate blood, which is processed to have the cells returned to the donors, and then the liquid portion of the blood, the plasma, <u>with its antibodies</u> can be given to someone suffering in the very early stages of the viral infection. These antibodies can bind to the virus, jump starting the immune system's response and helping to block the viruses.

This is used during treatment for a suspected rabies infection. In addition to giving a vaccine (that looks like surface antigens of the rabies virus), we also give antibodies to the rabies virus that were made in another animal, to try and help boost the patient's immune response to the rabies virus before it can settle into the brainstem and cause symptoms. As we mentioned before, if rabies is left untreated it is 100% fatal.

During the Ebola outbreak in 2014, two different experimental treatments were used. One included antibodies that were made in tobacco plants. When patients were given these antibodies, it helped their immune system target infected cells and virus particles. Not every patient that received this treatment survived, but it did seem to help some patients, as long as they also had supportive care like dialysis (for kidney failure), respirators (for lung failure), and fluids (to prevent dehydration from diarrhea).

In the early stages of the COVID-19 pandemic, survivor plasma was used as a tool to try and jumpstart the immune response of people in the early stages of the infection. A few months later, monoclonal antibodies created in the laboratory were used to do this, as well. Once the illness has progressed far enough to cause the patient to produce their own antibodies, adding survivor or monoclonal antibodies doesn't help, and can in fact make deadly inflammation worse. This is definitely an early-stage-only intervention.

Anti-inflammatory medicines

With a mild viral illness, like a cold, your symptoms are caused by your inflammation response as the immune system gears up to produce antibodies and destroy the invaders. These mild inflammation responses help white blood cells reach the invaders, and help your body activate your immune system. (In the meantime, you have a headache, congestion, or a fever.) This is why often the best way to treat an infection like this is to get lots of rest, take care of yourself, and take an anti-inflammatory like ibuprofen.

But sometimes, a viral infection is severe. In this case, the damage a virus does to a patient's organs is an indirect effect of the virus; the virus triggers an overwhelming immune response that causes inflammation. Inflammation can be a good thing in a limited way, encouraging and making it easier for the immune system to destroy invaders. But out-of-control inflammation can cause damage, like blood clots in critical organs, or fluid and scarring in the lungs (Acute Respiratory Distress Syndrome, ARDS). Sometimes health professionals have to use anti-inflammatory medicines to try and control inflammation. Sometimes the risk to major organs is so severe that the organs have to be supported, like extra oxygen or ventilation for lung damage, or **dialysis** (external waste removal from the blood) for kidney damage.

## HEY, YOU KNOW WHAT DOESN'T WORK? JUST ABOUT EVERYTHING IN THE MOVIES.

Some zombie stories have invented a virus that causes zombieism, and then the story comes up with a "vaccine" of an unrelated bacterial strain that "combats" the infection. First of all, that's not what a vaccine is. Go review if you need to. A vaccine is a dead, empty, or weakened form of a virus, bacterium, or parasite administered to elicit an immune response.

So, right there, we have a problem.

Then, infection with a bacterium does not prevent infection with a virus. Remember the virus is targeting your own cells. Simultaneous bacteria in your lungs or bladder are not going to stop those viruses. Viruses don't have a brain to "care" whether you are infected with something else at the same time. They just do their thing.

So, there's the other problem.

However, there *is* a really cool idea.

There are viruses that specifically infect bacteria cells. They are called bacteriophages. There is a bacteriophage that specifically infects the bacteria species *E. coli*. There's another bacteriophage type that specifically infects *S. aureus*.

So… *if* the zombie outbreak was caused by a bacterium… and we could figure out what species it was… and we could find a bacteriophage that specifically infected and killed that species of bacteria… now *that* might work.

But again, that would take years and years to figure out and then develop into a medicine.

# CHAPTER 9

## QUESTIONS/WORDSTEMS

1. Establishing 85% Herd immunity for the measles pathogen is a way of
   a. reducing the odds of contact between an infected and vulnerable patient
   b. saving 85% of people from the infection
   c. protecting the strongest immune system in a population
   d. ensuring everyone is immune to the pathogen at all times
   e. protecting medical workers from measles virus

2. What is a vaccine made of?

3. How does penicillin work to stop bacterial reproduction?

4. Antibiotic resistance means
   a. a bacterial cell is not affected by a specific antibiotic and keeps living and growing
   b. all bacteria are no longer killed by any antibiotics
   c. resistance is useless!
   d. a person becomes antibiotic resistant when they cannot tolerate antibiotics anymore
   e. a person becomes antibiotic resistant when you cannot use antibiotics to cure them anymore
   f. a person becomes antibiotic resistant when you cannot use antibiotics to treat their flu anymore

5. Explain how natural selection happens (for example, for a penicillin-resistant Staphylococcus aureus).

| WORDSTEMS | |
|---|---|
| vac- | empty |
| entero- | within, within the intestine |
| pyogen- | producing pus |
| pneumo- | lungs |
| anti- | against; opposed |
| compromise- | to put at risk |
| immun- | unburdened |
| -phage | eat |

# CHAPTER 9

## WORSHEET

## ANTIBIOTIC RESISTANCE

### OBJECTIVES

- To show how antibiotic resistance increases in a population if antibiotics are misused.
- To demonstrate the necessity for using antibiotics correctly for the prescribed length of time.

### INTRODUCTION

Among the most important challenges in the healthcare field is the proliferation of pathogens that are resistant to antibiotics. Misuse or overuse of antibiotics contributes to the increase in antibiotic resistant pathogens.

There has been a lot of research into length of antibiotic courses, to determine the shortest possible length of course needed to completely kill all bacteria. If you are being treated for an infection, best practice is to administer the antibiotics for the full length of time per your physician's orders. Feeling better, or an improvement in symptoms, does not always mean that the infection has completely gone.

In this lab we'll simulate an *Otitis Media* (inner ear) infection. Beads will represent *Streptococcus pneumoniae* strains. Many strains of *S. pneumoniae* have developed resistance to penicillin. Table 9.1 below shows the different strains of *S. pneumoniae* along with their level of antibiotic resistance.

**TABLE 9.1**  Resistance of Strains of *S. pneumoniae* to Penicillin

| Strain | Color | Level of Resistance | Penicillin Efficacy |
|---|---|---|---|
| Strain 1 | white | susceptible | 50%/ day |
| Strain 2 | blue | intermediate | 25%/ day |
| Strain 3 | red | resistant | 10%/ day |

To simulate administration of antibiotics, for each day on the medication students will remove the percent of each strain as listed above in Table 9.1. To simulate "skipping" medication, students will double bacterial numbers daily.

## MATERIALS

Colored beads
Two bowls

## PROCEDURE

1. Separate the assorted colored beads.

2. For the first experiment, simulate using antibiotics for 10 days.
   a. Place 80 white, 15 blue, and 5 red beads in the first bowl. Record the numbers in the first column in Table 9.2.
   b. Keep removing the correct percent of each colored bead for 10 "days."
   c. Place the beads you removed in the second bowl. Record the data in the first column of Table 9.2.
   d. For this experiment, we'll consider a 75% reduction in the total microbial count as an indicator of when the symptoms and signs (*e.g.* pain, fever) subside. Did this occur? Indicate on the table how long it takes for the symptoms and signs to subside.
   e. For this experiment, we'll consider a 95% reduction in microbial count as an indicator that the infection has been wiped out and the immune system can finish off the remaining pathogens. Did this occur? Indicate on the table how long it takes for the symptoms and signs to subside.
   f. Using the data from the first experiment, construct a scatterplot w/ line graph using MS Excel. Indicate on the graph symptoms and signs subsiding along with the infection controlled.

3. For the second experiment, students will simulate using antibiotics only until signs and symptoms subside.
   a. Place 80 white, 15 blue, and 5 red beads in the first bowl. Record the numbers in the second column in Table 9.2.
   b. Keep removing the correct percent of each colored bead until signs and symptoms subside.
   c. Place the beads you removed in the second bowl. Record the data in Table 9.2.
   d. For the two days following, simulate skipping your dose of antibiotics. Double the number of microorganisms each day. Record the numbers.
   e. On the third day following subsiding of signs and symptoms, resume antibiotics. Remember you have a total of 10 days' worth of doses. Mark on the table if/ when symptoms and signs subside and if/ when the infection is controlled.
   f. Using the data from the second experiment, construct a scatterplot w/ line graph using MS Excel. Indicate on the graph symptoms and signs subsiding along with the infection controlled.

# RESULTS

## TABLE 9.2

| Experiment 1: 10 day prescription | | | | | Experiment 2: Interruption of medication | | | | |
|---|---|---|---|---|---|---|---|---|---|
| Day | White | Blue | Red | Total | Day | White | Blue | Red | Total |
| Initial count | | | | | Initial count | | | | |
| Day 1 | | | | | Day 1 | | | | |
| Day 2 | | | | | Day 2 | | | | |
| Day 3 | | | | | Day 3 | | | | |
| Day 4 | | | | | Day 4 | | | | |
| Day 5 | | | | | Day 5 | | | | |
| Day 6 | | | | | Day 6 | | | | |
| Day 7 | | | | | Day 7 | | | | |
| Day 8 | | | | | Day 8 | | | | |
| Day 9 | | | | | Day 9 | | | | |
| Day 10 | | | | | Day 10 | | | | |
| | | | | | Day 11 | | | | |
| | | | | | Day 12 | | | | |

## QUESTIONS

1. For experiment one, how long did it take for the signs and symptoms to subside? How long did it take to get the infection under control?

   _____

   _____

   _____

2. For experiment two, did the signs and symptoms subside AFTER skipping doses then taking the rest of the doses? Did the infection ever get under control?

   _____

   _____

   _____

3. Using this information, discuss the importance of using the entire prescription as prescribed by a physician.

_____

_____

_____

_____

_____

_____

_____

4. What's the only course of action for someone who skipped doses as in experiment 2?

_____

_____

_____

_____

_____

_____

_____

5. Using MS Excel, construct a graph of the results. Your independent variable will be time (days) while your dependent variables will be total bacterial count. Mark on the graph the level of bacterial reduction needed to wipe out the immune system (95% reduction).

# CHAPTER 10

## PANDEMICS IN MODERN TIMES
### (OR...HOW I LEARNED TO STOP WORRYING AND LOVE THE VACCINE)

# INFLUENZA VIRUS

The flu is a respiratory illness you catch caused by influenza viruses. They infect the nose, throat, and lungs. It can cause mild to severe illness and can even lead to death.

Flu viruses spread by droplets made when people with flu cough, sneeze or talk. These droplets can land in the mouths or noses of people who are nearby. A person might also get flu by **fomites** or touching a surface or object that has flu virus on it and then touching their own mouth, eyes, or nose. You may be able to pass on the flu to someone else before you know you are sick and while you are sick. Unfortunately, healthy adults might infect others beginning one day before symptoms even develop so you don't know you are sick. Sharing infection with your friends can also occur up to a week after becoming sick. Young children and people with weakened immune systems might be able to share infection with others for an even longer time. The best way to prevent the flu is by getting a flu vaccine each year.

There are 2 important types of flu viruses that infect humans: influenza A and influenza B. Influenza A viruses can infect any mammal. Wild birds are the primary natural **reservoir** for all subtypes of influenza A viruses and are probably the source of influenza A viruses that infect all other animals. Most Influenza A viruses cause asymptomatic (without symptoms) or mild infection in birds (avian flu). They also spread in pigs (swine flu) and humans. Influenza B viruses are spread only between humans.

Not only do pigs get symptoms, just like humans, such as cough, fever, and a runny nose, they also play a very special role in the creation of new flu viruses. They can be infected with human, avian, and swine flu viruses and even at the same time. If this happens, it is possible for the genes of these viruses to mix and create a new virus. This is why pigs are often called the "mixing vessels" for creating new flu viruses.

# INFLUENZA A CYCLE

**FIGURE 10.1**                                                                 Influenza A Virus.

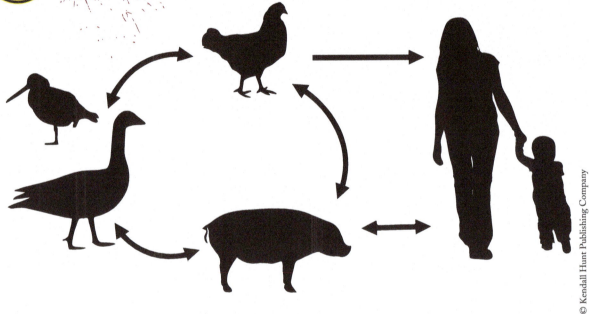

© Kendall Hunt Publishing Company

Influenza A viruses are divided into subtypes based on two proteins on the surface of the virus, hemagglutinin (H) and neuraminidase (N). The combination of H & N proteins is dictated by the rearrangement of genes in the influenza virus. **Genetic or antigenic drift** (minor changes) and **genetic or antigenic shift** (major changes) are responsible for the differences in subtypes and strains. Genetic drift occurs when small nucleotide substitutions occur resulting in slight mutations in the H & N surface proteins. Genetic shift occurs when large gene deletions, substitutions, or **recombination** (think pigs!) take place and create an entirely novel virus never seen before. The immune response and previous flu vaccines for viruses that have drifted may still provide some cross-protection. When a novel virus is created from a shift, there is no cross-protection from previous infections or vaccines, since it has never circulated before. Flu is unpredictable and how severe it is can vary widely from one season to the next depending on many things, including what flu viruses are spreading, how much flu vaccine is available, when vaccine is available, how many people get vaccinated, and how well the flu vaccine is matched to flu viruses that are causing illness.

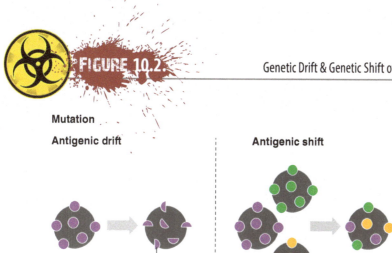

FIGURE 10.2.

Genetic Drift & Genetic Shift of Influenza.

**Mutation**

**Antigenic drift**

**Antigenic shift**

Small
mutations

New
strain

There are 18 different hemagglutinin subtypes, H1 through H18, and 11 different neuraminidase subtypes, N1 through N11. The naming of different flu viruses comes from whether it is influenza A or B, the host of origin (swine, horse, etc., or blank if human), geographical region where first isolated (Denver, Taiwan, etc.,), strain number (7, 15, etc.,), and year of isolation (57, 2009, etc.,). For influenza A viruses, the hemagglutinin and neuraminidase antigen descriptions are in parentheses (e.g., (H1N1), (H5N1)). Strains in the 2022-2023 quadrivalent flu vaccine: A/Victoria/2570/2019 (H1N1)pdm09; A/Darwin/9/2021(H3N2); B/Austria/1359417/2021; and B/Phuket/3073/2013.

**TABLE 10.1.**

| Influenza A Subtypes that Cause Human Infection | |
|---|---|
| Avian | H5N1 |
| | H7N3 |
| | H7N7 |
| | H7N9 |
| | H9N2 |
| | H10N8 |
| Swine | H1N1 |
| | H1N2 |
| | H3N2 |
| Most Common | H1N1 |
| | H3N2 |

A novel flu pandemic is a global outbreak of a **novel** (meaning new) influenza A virus (only influenza A causes pandemics) in people that is very different from current and recently circulated seasonal influenza A viruses. Novel pandemic flu viruses spread in the same way as seasonal flu, but a pandemic flu virus will

# 1918 SPANISH INFLUENZA AND 2009 SWINE FLU PANDEMICS

One third of the world population was hit with the deadliest pandemic in recorded history from January 1918 to December 1920. Up to 5% of the world's population (by some estimates 100 million) died from what was known as the Spanish Flu pandemic. This is called Guillain-Barre syndrome (GBS). This is a true concern of medical personnel, but the risk of GBS is much higher as a result of actually catching a virus, compared to the risk from a vaccine preventing that viral infection.

The causative agent was a strain of influenza virus now known as H1N1. H1N1 causes an overreaction of the body's immune system. As a result, while most strains of influenza kill immunocompromised individuals such as the elderly, very young, or otherwise weakened, H1N1 kills healthy young adults, via strong immune reaction. The mortality rate of H1N1 approached 20%, as opposed to 0.1% for typical flu strains.

Fast forward to 2009 and the Swine Flu epidemic. Again, the culprit was H1N1. The current world population at 7 billion+ meant that 350 million could have succumbed to the disease. This time, however, the CDC was ready with both a live (attenuated) and inactivated vaccine. The U.S. government and CDC highly encouraged everyone to get vaccinated against 2009 H1N1.

Unfortunately, the anti-vaxxers were also ready. TV and radio personalities actively pushed against vaccinations.

"Who put the notion that you gotta have this shot, or this nasal spray—whatever the hell the vaccine is—whoever the hell put in your head the notion that you gotta do it? Government did. The Obama government, to be specific.
—Rush Limbaugh [The Rush Limbaugh Show, 10/7/09]

Luckily, in spite of opposition the CDC was able to ensure that the 2009 H1N1 pandemic never reached the deadly numbers from the 1918 pandemic. But make no mistake. Influential people convinced Americans not to vaccinate, and Americans died.

**FIGURE 10.3.**  Temporary housing for Spanish Flu victims, Lawrence, MS 1919).

Source: Library of Congress

likely infect more people because few people have immunity to the pandemic new strain. Underlying health conditions of the population may play a more or less important role than the difference of a new flu strain. Novel flu pandemics happen rarely. Four novel flu pandemics have happened in the past 100 years (i.e., 1918, 1957, 1968, 2009), but another one could happen any time.

Underlying health conditions did not play a role in the spread of these infections.

## 1957 Asian Flu (H2N2) Pandemic

Although the 1957 pandemic was not as deadly as the 1918 Spanish Flu, over a million people died worldwide, with over 100,000 of those in the United States. The novel influenza virus (H2N2) originated in Asia in early 1957. The H and N antigens were unlike any previously circulating in humans. Except for persons >70 years old, people were faced with a virus never seen before. The virus was not more virulent than previous subtypes, but people with underlying health conditions of the heart or lungs experienced higher mortality. Deaths of previously healthy persons were also not uncommon. Rheumatic heart disease was the most common preexisting condition and women in the third trimester of pregnancy were most at risk. This was the first time the rapid global spread of a modern influenza virus was available for laboratory investigation and vaccine response. Since the Asian flu virus was novel and the population of the world was immunologically naïve, more vaccine was needed to ramp up a primary antibody response in the population. The Asian flu virus disappeared 11 years after it was identified. It was replaced by the Hong Kong (H3N2) subtype.

## 1968 Hong Kong Flu (H3N2) Pandemic

Another rapidly spreading influenza A virus (H3N2) was discovered in Asia in 1968. Many cases, people with laboratory confirmed isolation of pathogen, were identified in Hong Kong, and got the Western world's attention. Major differences in the pattern of illness and death were noted from the 1918 Spanish flu and 1957 Asian flu pandemics. In some Asian countries the epidemics were small; however, in the United States the illness and death rates were high. This was also contrary to Western Europe where cases of disease were high, but death did not happen as often until later in the pandemic. The N antigen for this virus is the same as the 1957 Asian flu but the H antigen was new. There was reduced illness in people who had been sick or vaccinated for the 1957 Asian flu. Greater disease severity occurred in this pandemic than the 1957 Asian flu pandemic and the virus was able to infect more species than just humans.

## Prepped for Flu – Pan Flu Plans

Having experienced 4 influenza pandemics in less than 100 years, and living with the uncertainty of when a novel pandemic strain of flu might appear, governments developed novel pandemic influenza or "pan flu" plans. Communities, institutions, businesses, and schools were advised to do the same. The purpose of a pan flu plan is to provide guidance in preparing for, identifying, and responding to novel pandemic flu. The plan provides a framework for pan flu preparedness and response activities and serves as a foundation for further planning, drills, and emergency preparedness activities.

FIGURE 10.4.

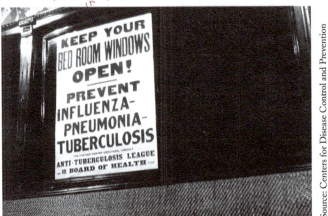

Source: Centers for Disease Control and Prevention

It is important to remember that pan flu plans were developed specifically to stop and prevent the spread of influenza viruses. Influenza viruses are transmitted by fomites and person-to-person by large droplets. Because the large droplets are heavy and fall to the ground before they travel very far, **social distancing** or creating space between people in public settings is very important to reduce influenza spread. This is why wearing a mask if you are sick with the flu will help reduce spread; because the large droplets will be caught by the mask. Healthy people are usually not recommended to wear masks to stop **transmission** or spread of a novel flu in a pandemic, since they can't inhale the large droplets if they are socially distanced (at least 6 feet apart) and washing their hands helps reduce fomite spread. Healthy people wearing masks are only helpful when a respiratory disease is transmitted by aerosol (like the measles virus).

Critical pieces of the pan flu plan, preparedness and response, are having a plan to disseminate timely, consistent, and accurate information to the public. Other pieces include pharmaceutical and non-pharmaceutical prevention strategies and interventions. Pharmaceutical prevention strategies and interventions include seasonal and novel influenza vaccinations. Seasonal influenza vaccination may or may not offer some level of preventive protection against a novel flu virus. To create a pandemic-specific flu vaccine could take six months or more from the time a candidate is placed into production. Once the pandemic strain flu vaccine is available, it may take five months to produce an adequate supply of vaccine for the entire U.S. population.

Non-pharmaceutical preventions for flu include social distancing or creating ways to increase distance between people in settings where people commonly come into close contact with one another. Cancelling events that involve large gatherings of people, encouraging surface cleaning of high touch surfaces, focus cleaning efforts in common areas on a more regular basis, and communicating with the community about every day preventive actions are part of these preventions. Staying home when sick, covering coughs and sneezes with a tissue or sleeve, washing hands with soap and water, or using hand sanitizer are all important. Non-pharmaceutical interventions are the primary way to control disease until supplies of vaccines and/ or antiviral medications are available. Over the course of a pandemic, up to 50% of the workforce may be absent due to illness, caretaking responsibilities, fear of contagion, loss of public transportation, or public health control measures.

# CORONAVIRUSES

Coronaviruses (CoV) are a family of viruses that can cause respiratory infection. Coronaviruses are named for the crown-like spikes that surround their surface. Coronaviruses can cause mild infections in humans, such as the common cold, or deadly infections, such as SARS, MERS, and COVID-19. Some coronaviruses are known to occur only in animals. A coronavirus that infects animals can evolve to infect people, and then start to spread from person-to-person. Such an infection that spreads from animals to people is said to be a "spillover" zoonosis. Because humans have not built up any immunity to these new viruses, the illness they cause can be severe.

## SARS 2003 Pandemic

Severe acute respiratory syndrome (SARS) is a viral respiratory illness caused by a coronavirus, called SARS-associated coronavirus (SARS-CoV). SARS originally appeared from an infected palm civet and/or other animals that are available and eaten from Asia food markets that we don't eat here in America. SARS was first reported in Asia in February 2003. Over the next few months, the illness spread to North America, South America, Europe, and Asia before the outbreak was contained. According to the World Health Organization (WHO), a total of 8,098 people worldwide became sick with SARS during the 2003 outbreak. Of these, 774 died. In the U.S., only eight people had SARS-CoV virus confirmed by laboratory isolation methods, and all had a history of travel parts of the world with SARS infection. SARS disease did not spread person-to-person in the U.S, although it did in other parts of North America. SARS infection was more common in men. SARS infection was fatal for 1 in 10 patients with age being a key factor in risk of mortality. The risk fatality increased to ~50% of cases over 60 years old. Only a small percentage of recovered cases had long-term effects from their illness, including depression or anxiety, cough, shortness of breath, chronic lung disease, or kidney disease.

The incubation period, the amount of time the pathogen needs to duplicate in a host to cause disease, for SARS disease is between two and 10 days. Interestingly though, SARS-CoV becomes more contagious during the last three days of infection. SARS usually began with a high fever (>100.4°F [>38.0°C]). Other symptoms included headache, an overall feeling of discomfort, and body aches. Some people also had mild respiratory symptoms from the beginning. About 10%-20% of patients had diarrhea. Between 2 to 7 days, SARS cases could develop a dry cough. Most patients developed pneumonia.

The main way that SARS-CoV was transmitted was close person-to-person contact through respiratory droplets when an infected person coughed or sneezed. The virus also spreads by fomites. In addition, it is possible that SARS-CoV might have spread more broadly through the air (airborne spread) or by other ways that were not identified. The CDC recommended that patients with SARS infection receive the same treatment that would be used for any serious community-acquired atypical pneumonia.

Although there have been no cases of SARS infection anywhere in the world since 2004, preventing the spread of this illness was like preventing any large-droplet viral respiratory infection. Namely avoiding close contact with affected individuals, washing hands with soap and water, and encouraging people with viral respiratory infections to cover their mouth when coughing or sneezing. In 2012, the National Select Agent Registry Program published a final rule declaring SARS-CoV a select agent. Meaning SARS-CoV has the potential to pose a severe threat to public health and safety should it resurface.

# COVID-19 THE FIRST YEAR ORDER OF EVENTS

This is a brief introduction into how the COVID-19 pandemic started. There are many more details in the timeline than this passage captures.

**December 2020**
- FDA issues an EUA for the Pfizer-BioNTech's and Moderna's COVID-19 vaccine
- Sandra Lindsay, a nurse in New York, becomes the first American to receive a COVID-19 vaccine
- The U.K. announces the detection of a new and more contagious COVID-19 variant, B.1.1.7

**September 2020**
- Pfizer BioNTech and Johnson & Johnson begin phase 3 clinical trials of COVID-19 vaccine

**August 2020**
- COVID-19 positive people are infectious to others for up to 10 days after symptoms first appear
- First documented case of reinfection in U.S.

**July 2020**
- WHO announces that COVID-19 disease can be transmitted through the air and is likely spread by asymptomatic individuals

**April 2020**
- Masks recommended outside of home in U.S.
- Operation Warp speed initiated

In December of 2019, the WHO noticed a media statement about a cluster of atypical viral pneumonia cases in Wuhan, China. By early January 2020, China was sharing information on the 44 cases with the WHO to help identify the cause of the illnesses. The WHO shared information on the cluster and advised other nations to be aware of similar infections of unknown **etiology**, take precautions to reduce the risk of transmission, and use surveillance for influenza and severe acute respiratory infections. A novel coronavirus not previously identified in humans was isolated soon after in China. As the RNA from the virus was sequenced, China reported the first death from the outbreak.

### December 2019
- A cluster of atypical viral pneumonia cases of unknown etiology are reported to WHO

### January 2020
- Novel coronavirus was identified and the RNA from the virus was sequenced
- First death in China
- First lab confirmed case outside of China (Thailand)
- The first advice to use medical masks in conjunction with other infection control measures came from the WHO for symptomatic people
- Confirmation of person-to-person transmission; WHO Director-General declared the novel coronavirus outbreak a PHEIC

### February 2020
- WHO advises implementing large-scale public health measures
- Effective prevention methods based on pan flu plans and large droplet spread

### March 2020
- COVID-19 outbreak declared a pandemic by WHO
- U.S. States implement shutdowns
- Moderna Therapeutics begin the first human vaccine trials against COVID-19
- Critical shortage of ventilators, hospital beds, blood products, PPE, and body bags in New York City

By January 13, 2020, Thailand had identified the first imported case of a lab-confirmed, novel coronavirus infection, being the first case identified outside of Wuhan, China. Two days later the Japanese Ministry of Health confirmed a case of novel coronavirus in a person who had travelled to Wuhan. A WHO risk assessment indicated a need to investigate the presence of human-to-human transmission, modes of transmission, common source of exposure, and the presence of asymptomatic or mildly symptomatic cases that were undetected. There were forty-one confirmed cases worldwide as of the next day, January 14.

The single-stranded RNA genome of the SARS-CoV-2 virus was published on GenBank® (the NIH genetic sequence database) for public access on January 17, only six days after a draft genome was released. On January 19, 2020, the WHO Western Pacific Regional Office (WHO/WPRO) tweeted that the latest information indicated mild person-to-person transmission of the novel coronavirus. Within 2 days, transmission to healthcare workers provided support for this. The first confirmed case of novel coronavirus in the U.S. was then identified in Washington State. At this point, the WHO Director-General did not identify this as a public health emergency of international concern (PHEIC) probably due to mostly localized spread. January 24, 2020, Wuhan was placed under lockdown and the first cases were identified in Europe with travel history to Wuhan. The Director of the Pan American Health Organization (PAHO) urged countries in the Americas to be prepared in case of receiving travelers from countries where there was ongoing transmission of novel coronavirus cases. United Arab Emirates announced the first confirmed case of the novel coronavirus in the Eastern Mediterranean region on January 29, 2020. There were 4,690 confirmed cases worldwide.

On January 29, 2020, the first advice to use medical masks in conjunction with other infection control measures (hand washing, social distancing, etc.,) came from the WHO for symptomatic people. Building on the pan flu plans previously established, and assuming SARS-CoV-2 was transmitted similarly through large droplet spread, a warning about wearing a mask incorrectly was advised stating that it could hamper the effectiveness of reducing the risk of transmission and that healthy people should not wear a mask. Cloth (e.g., cotton or gauze) masks were not recommended under any circumstance. January 30, 2020, the Director-General declared the novel coronavirus outbreak a PHEIC, WHO's highest level of alarm. The same day the CDC confirmed that the SARS-CoV-2 virus had spread between two people in Illinois with no history of recent travel. This was the first recorded instance of person-to-person spread in the U.S. On

> February 4, 2020
> "We have a window of opportunity. While 99% of cases are in China, in the rest of the world we only have 176 cases."
> WHO Director-General

On February 9, 2020, WHO reported 37,558 confirmed cases worldwide. February 11, 2020, the WHO announced the name of the novel coronavirus as Coronavirus Disease 2019 or COVID-19. At this point the main clinical signs and symptoms reported in people included fever, coughing, difficulty breathing, and chest radiographs showing bilateral lung infiltrates. Person-to-person transmission was confirmed in Wuhan and other locations in China, and internationally. Using Ebola and 2009 Swine flu (H1N1) outbreak guidance, the WHO released guidance for mass gatherings regarding COVID-19 on February 14, 2020.

After assessing the situation in Wuhan, the WHO-China Joint Mission on COVID-19 held a press conference to report the main findings on February 24, 2020. They warned that most of the world was not prepared to implement measures to contain COVID-19 and that to reduce infection and death, implementing large-scale public health measures were necessary. Including case detection and isolation, contact tracing and quarantining, and community engagement. Confirmed cases started being reported on the African continent. Moderna delivered the first mRNA vaccine to the National Institutes of Health for testing.

As of February 27, 2020, the most effective preventive measures in the community, based on the pan flu plans and assuming large droplet spread, included performing hand hygiene frequently with an alcohol-based hand rub if hands were not visibly dirty or with soap and water if hands were dirty; avoiding touching the eyes, nose, and mouth; practicing respiratory hygiene by coughing or sneezing into a bent elbow or tissue and then immediately disposing of the tissue; wearing a medical mask if you had respiratory symptoms and performing hand hygiene after disposing of the mask; and maintaining social distance (a minimum of 6 feet) from individuals with respiratory symptoms. On February 29, 2020, there were 85,403 confirmed cases worldwide and 2,924 deaths.

March 7, 2020, the number of confirmed COVID-19 cases was over 100,000 globally. March 11, 2020, WHO characterized COVID-19 as a pandemic. March 12, 2020, U.S. states started the first COVID-19 restrictions and shut-downs. March 17, 2020, the first Moderna mRNA vaccine started testing in human subjects.; Pfizer starts collaborations with BioNTech to develop an mRNA vaccine. March 27, 2020, the U.S. government enacted the Coronavirus Aid, Relief, and Economic Security (CARES) act. The act included funding of $1,200 per adult (more for families with children), expanded unemployment benefits, forgivable small business loans, loans to major industries and corporations, and expanded funding to state and local governments in response to the economic crisis caused by COVID-19. The global confirmed case count was 750,890 with 36,405 deaths on March 31, 2020.

April 3, 2020, CDC announced new mask wearing guidelines and recommended that all people wear a mask when outside of the home. April 4, 2020, over 1 million cases of COVID-19 had been identified globally. Because of the ability of other coronaviruses to be transmitted by fomites, the WHO continued with the measures originally designed for large droplet spread and did not recommend that healthy people wear masks to prevent transmission of SARS-CoV02 virus.

By April 13, 2020, most states in the U.S. reported widespread cases of COVID-19 disease.

April 20, 2020, shortage of personal protective equipment (PPE) like gowns, eye shields, masks, and even body bags, became dire, particularly in New York State where case counts were higher than some countries. April 22, 2020, two cats tested positive for the SARS-CoV-2 virus. April 30, 2020, **Operation Warp Speed**, an initiative to produce a vaccine against the SARS-CoV-2 virus as quickly as possible, is launched amidst 3,090,445 confirmed global cases and 217,769 deaths.

May 1, 2020, WHO declared that the global COVID-19 pandemic remained a PHEIC.

May 28, 2020, the recorded death toll from COVID-19 in the U.S. surpassed 100,000 while global confirmed cases reached 5,701,337 with 357,688 deaths.

June 10, 2020, the number of confirmed COVID-19 cases in the U.S. surpassed 2 million. Global confirmed cases: 10,185,374; deaths: 503,862.

July 7, 2020, the number of confirmed COVID-19 cases in the U.S. surpassed 3 million. July 9, 2020, WHO announced that SARS-CoV-2 virus was transmitted through the air (aerosolized) and was likely being spread by asymptomatic individuals. Recommendations were changed as understanding of the pathogen changed, as would be expected in any pandemic response. This is a benefit of changing knowledge based on science. July 14, 2020, CDC again called on all people to wear cloth face masks when leaving their homes to prevent the spread of COVID-19. July 16, 2020, the U.S. reports a record number of COVID-19 infections reported in a single day (75,600). July 31, 2020, 17,106,007 confirmed cases were documented globally with 668,910 deaths.

August 17, 2020, COVID-19 became the 3rd leading cause of death in the U.S. Deaths from COVID-19 exceeded 1,000 per day and nationwide confirmed cases exceeded 5.4 million. August 26, 2020, FDA issued an **Emergency Use Authorization (EUA)** for Abbott's BinaxNOW COVID-19 test kit, a rapid antigen test that could detect a COVID-19 infection in 15 minutes using the same technology as a flu test. August 24, 2020, the University of Hong Kong confirmed the first documented case of COVID-19 reinfection. Four days later, the first documented case of reinfection was confirmed in the U.S.

September 22, 2020, the reported death toll in the U.S from COVID-19 surpassed 200,000. September 28, 2020, 10 months into the pandemic, the reported death toll from COVID-19 reached more than 1 million worldwide.

October 5, 2020, CDC updated its "How COVID-19 is Spread" guidelines acknowledging the potential for the airborne spread of the SARS-CoV-2 virus even when 6 feet of social distance was maintained if the area is poorly ventilated or enclosed and activities occurred that required heavy breathing (singing, exercise, etc.,). Recommendations were adapted when new scientific information came to light. These new recommendations no longer followed the pan flu plans. On October 18, 2020, over 2.4 million confirmed cases and 36,000 deaths occurred in the previous week making a total of over 40 million confirmed cases worldwide with 1.1 million deaths.

By mid-November both Moderna and Pfizer-BioNTech's COVID-19 vaccines were found to be 94% and 95% effective, respectively. Headed into the U.S. Thanksgiving holiday on November 26, 2020, global confirmed cases exceeded 57.8 million with 1.3 million deaths.

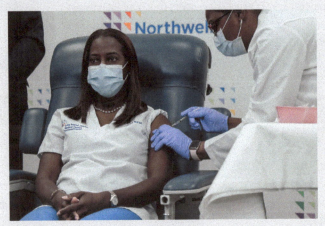

Xinhua /Alamy Stock Photo

December 11, 2020, FDA issued an EUA for the Pfizer-BioNTech COVID-19 vaccine. Sandra Lindsay, a critical care nurse at Long Island Jewish medical Center, received the first Pfizer-BioNTech's COVID-19 vaccine outside of clinical trials in the U.S.

[December 14, 2020, the recorded death toll from COVID-19 in the U.S surpassed 300,000 and the U.K. announced the detection of a new and more contagious COVID-19 variant, B.1.1.7. December 18, 2020, FDA issued an EUA for the Moderna COVID-19 vaccine. December 29, 2020, the first case of the COVID-19 B.1.1.7 /"Alpha" variant was detected in the U.S. by the Colorado Department of Health.

December 31, 2020, one year after the first reported case of COVID-19, 4 million confirmed cases were being reported each week worldwide. Since vaccine EUA 2 weeks earlier to Moderna and Pfizer-BioNTech, 2.8 million people in the U.S. had received a COVID-19 vaccine dose.

## COVID-19 PANDEMIC

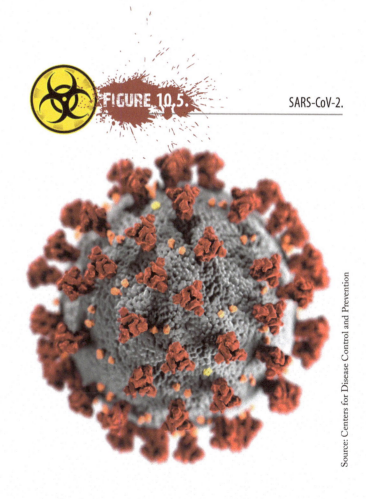

FIGURE 10.5.  SARS-CoV-2.

**COVID-19** infection is a viral respiratory illness caused by a coronavirus, SARS-CoV-2. The naming convention for this disease/pathogen combination is like others, such as plague, which is caused by the bacterium *Yersinia pestis*. COVID-19 disease symptoms include fever, cough, and shortness of breath, which may or may not be accompanied by gastrointestinal-specific symptoms. Scientific study shows that SARS-CoV-2 is transmitted by respiratory droplets and as an aerosol. To prevent transmission, social distancing between people who do not live in the same household, mask wearing of infected and non-infected people, and frequent hand washing should be practiced. These health behaviors reduce the chance of pathogen spread by an infected person sneezing, coughing, or talking. The incubation period for COVID-19 is between 2-14 days. Unfortunately, COVID-19 is another disease that has a long **presymptomatic transmission** period, or transmission of infection before symptoms start, driving home the importance of practicing control measures even by people who assume they are currently healthy. (This is another big difference from flu virus, which rarely spreads presymptomatically and if it does, has only a day or two to do so before symptoms start).

It turns out that SARS-CoV-2 viral RNA can be found on surfaces, in drinking-water, and sewage/wastewater, but finding an entire infectious virus in these contexts is not very likely. Unlike influenza and other coronaviruses, the risk of contracting COVID-19 disease from fomites is low (1 in 10,000). In fact, after 3 days, even if surfaces have not been cleaned, the risk of fomite transmission is minimal. Disinfection is still recommended in indoor community settings where there has been a suspected or confirmed case of COVID-19 disease within the previous 24 hours.

Isolation and quarantine of cases and close contacts were heavily used in this pandemic. **Isolation** is used for people with COVID-19 diagnosed by symptoms or laboratory-confirmation. They are directed to stay at home for a 10-day minimum from symptom onset. If their fever hasn't been gone for at least 24 hours, without the use of fever-reducing medications, and with improvement of other symptoms, their isolation may be extended. **Quarantine** is used to separate and restrict the movement of people exposed (or potentially exposed - contacts) to a COVID-19 confirmed case. Historically, quarantine is 2 weeks, and this is what we did in the COVID-19 pandemic. Quarantine in contacts starts the date from the last exposure with the confirmed case and continues for 14 days.

During isolation, a case is interviewed to find out where they may have been potentially exposed and determine close contacts who may have been exposed to the disease during their infectious period. In the case of COVID-19, close contact is someone who was within 6 feet of a confirmed case for at least 15 minutes over a 24-hour period, while the case was infectious. Contacts are then interviewed. The main goal of a contact interview is to prevent others from getting or spreading the disease by providing information, education, and support. Contacts are advised to self-quarantine for 14 days from the date of their last exposure to the active case. The contact's date of last exposure is determined by the last time they were exposed to the case during their infectious period.

It became clear that pan flu measures to stop disease spread was not going to be sufficient to stop the Covid-19 wave, so a mitigation strategy to prevent healthcare facilities and providers from being overwhelmed was enacted. **Flattening the curve** was encouraged through isolation and quarantine, so people with the pathogen (who may have been asymptomatic or had mild symptoms) didn't continue to go out in public and spread disease. The focus was not preventing the spread of disease overall; it was to lessen the daily cases of disease so the healthcare system could care for COVID-19 patients and maintain some sort of normalcy for other patients, especially those who have emergencies. Despite the effort to flatten the curve, the healthcare system in the U.S. was overwhelmed within the first month of disease transmission.

Actions to help flatten the curve:

- Wash your hands regularly for a minimum of 20 seconds
- Avoid touching your eyes, nose, and face
- Cover your cough and sneeze with a tissue or the inside of your elbow to prevent droplet transmission
- Wear a mask if you are sick to transmission
- STAY HOME! Social distancing means avoiding close contact with others

FIGURE 10.6. COVID-19 Flatten the Curve.

## Flatten and shrink the curve

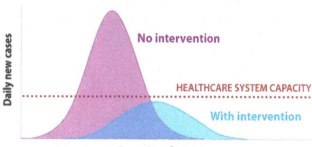

# CONTAINMENT *VERSUS* MITIGATION STRATEGIES

Containment and mitigation strategies are used at different timepoints in an outbreak or epidemic to reduce transmission and allow for proper allocation of resources.

- **Containment strategies** – used to prevent transmission of infection, usually at the very beginning of an outbreak or epidemic
  - Tracking the close contacts of confirmed cases, quarantining those who have had a known exposure to a case (but are not known to be infected yet), and isolating those with infection to prevent further transmission are containment classics
  - Testing is critically important during containment because we must know who is infected and/or exposed so they can be isolated to prevent further spread of infection
- **Mitigation strategies** – used to redirect resources to where they are of most help assuming everyone in the community is infected
  - Imposing social distancing policies, preparing hospitals and communities for a surge of patients, and working with schools and businesses on closure policies to reduce transmission (flatten the curve) are mitigation classics
  - Testing is not important, except when warranted for individual clinical decisions

In a situation like the COVID-19 pandemic, the only hope for actually stopping the spread of disease is either prevention of spread (think lockdowns and containment) or herd immunity. Remember herd immunity from chapter 9 means almost everyone has encountered the pathogen, either through natural course of disease or vaccination. Herd immunity means a contagious person encountering someone who has not had contact with the disease is very unlikely, so the spread stops. The amount of herd immunity required is specific to each pathogen based on how infectious it is and how easily it is transmitted. We use herd immunity to control a variety of contagious diseases, including influenza, measles, mumps, rotavirus, and pneumococcal disease.

FIGURE 10.7.     Herd Immunity.

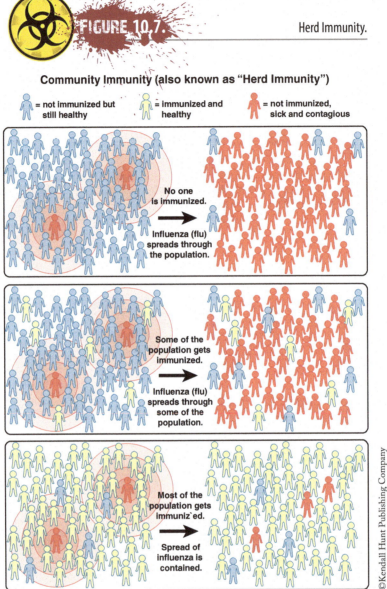

**Community Immunity (also known as "Herd Immunity")**

= not immunized but still healthy    = immunized and healthy    = not immunized, sick and contagious

No one is immunized.

Influenza (flu) spreads through the population.

Some of the population gets immunized.

Influenza (flu) spreads through some of the population.

Most of the population gets immuniz`ed.

Spread of influenza is contained.

©Kendall Hunt Publishing Company

# PUBLIC HEALTH MEASURES

The silver lining of the doom and gloom associated with pandemics is that Public Health in the U.S. has already implemented measures to prevent and/or mitigate the spread of diseases! These measures are things that we may not pay attention to because we see them every day without thinking about them. We take for granted that we are being looked after even when we aren't paying attention. Drinkable water, unpolluted air, and vaccinated populations are examples of public health measures briefly described below.

In the U.S. you can expect access to safe drinkable water, thanks Public Health! One of the ways we ensure access to safe water is by keeping sewage away from drinking water supplies. We have the infrastructure (sewers, toilets) to carry our waste away from water supplies. We also have the knowledge and ability to treat the water so it can be cleaned, recycled, and used again. Once the water is safe and clean, we carefully transport and store water to prevent contamination. By removing sewage contamination from water, we have eliminated diseases associated with unsafe water like cholera, typhoid, and polio. In other parts of the world, these pathogens still circulate from sewage into drinking water and cause many cases of illness and death, especially in young children.

Like clean water, air quality impacts more than just our health. There are many public health measures that we can implement and engage in to improve air quality. Reducing vehicle emissions through car inspections, switching to use unleaded gasoline, educating on the risks of indoor air pollution from indoor cooking over a wood fire, and evacuating residents in wildfire areas or providing clean air shelters are examples.

FIGURE 10.8

Public Health Measures.

Source: Felix Mizioznikov

New York City with urban smog; clean air regulations reduce respiratory disease.

Source: Maarten Zeehandelaar

Sewage treatment and water sanitation systems reduce disease spread.

Source: Dmitry Kalinovsky

Municipal garbage collection reduces disease spread.

# Vaccinations

Vaccination is arguably one of the most important measures that public health has in its toolbox. Through vaccination we have significantly reduced the number of infectious diseases of childhood and beyond. Vaccination is usually used as a prevention method to decrease the burden of disease in a population (like for measles, mumps, and rubella); however, in cases of influenza and COVID-19 disease, vaccination is also used to lessen the symptoms of disease in individuals. The immune system can then have time to fight off the pathogen, and we can lower the burden of disease on the healthcare system. Vaccinations are important to achieve both!

FIGURE 10.9.                                                                Vaccine Development Timeline.

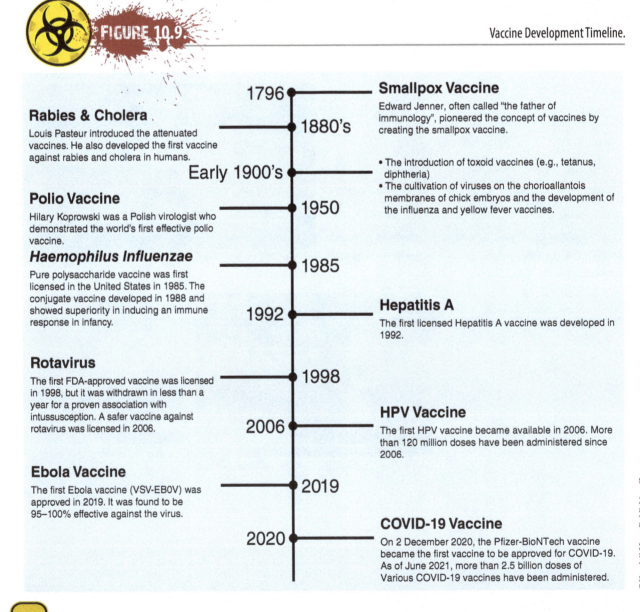

**Rabies & Cholera**
Louis Pasteur introduced the attenuated vaccines. He also developed the first vaccine against rabies and cholera in humans.

**Polio Vaccine**
Hilary Koprowski was a Polish virologist who demonstrated the world's first effective polio vaccine.

*Haemophilus Influenzae*
Pure polysaccharide vaccine was first licensed in the United States in 1985. The conjugate vaccine developed in 1988 and showed superiority in inducing an immune response in infancy.

**Rotavirus**
The first FDA-approved vaccine was licensed in 1998, but it was withdrawn in less than a year for a proven association with intussusception. A safer vaccine against rotavirus was licensed in 2006.

**Ebola Vaccine**
The first Ebola vaccine (VSV-EBOV) was approved in 2019. It was found to be 95–100% effective against the virus.

1796
1880's
Early 1900's
1950
1985
1992
1998
2006
2019
2020

**Smallpox Vaccine**
Edward Jenner, often called "the father of immunology", pioneered the concept of vaccines by creating the smallpox vaccine.

- The introduction of toxoid vaccines (e.g., tetanus, diphtheria)
- The cultivation of viruses on the chorioallantois membranes of chick embryos and the development of the influenza and yellow fever vaccines.

**Hepatitis A**
The first licensed Hepatitis A vaccine was developed in 1992.

**HPV Vaccine**
The first HPV vaccine became available in 2006. More than 120 million doses have been administered since 2006.

**COVID-19 Vaccine**
On 2 December 2020, the Pfizer-BioNTech vaccine became the first vaccine to be approved for COVID-19. As of June 2021, more than 2.5 billion doses of Various COVID-19 vaccines have been administered.

From the first vaccine developed in 1796 against smallpox to the latest COVID-19 vaccine, whether an inactivated form of the entire pathogen or a single piece of the genetic material, there have been several vaccines that have reduced the burden of disease, decreasing associated morbidity and mortality.

## SMALLPOX VACCINATION

Smallpox is a disease that used to spread easily and had a 30% mortality rate (even higher in infants). The disease was spread by droplets when coughing or sneezing and by fomites when pus or scabs from the pox blisters got on an inanimate object and was touched by another. Smallpox had such an impact on the human population that anecdotally families wouldn't count their children as part of the family until they had survived the smallpox infection! Yikes, the severity of this disease was no joke. Luckily, through vaccination, smallpox is the only human disease that has ever been eradicated, as it has no non-human hosts or reservoirs, and we were able to vaccinate enough of the population to reach global herd immunity.

FIGURE 10.10. Smallpox Infection.

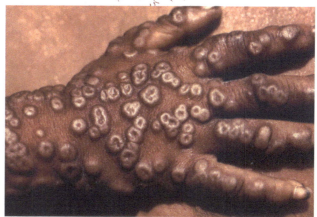

Source: Centers for Disease Control and Prevention

The Global Smallpox Eradication Program was started by the WHO in 1959. Although millions of vaccine doses were given during the first years of the campaign, 7 years later there were still outbreaks of disease scattered around the world. The Intensified Eradication Program started in 1967 to increase the push for global vaccination. This second program proved highly successful. Many countries were able to make a smallpox vaccine by this time,

## ANTI-VAXXERS VACCINE CONCERN: TOO MANY TOO SOON!

A concern for some anti-vaxxers is the number of vaccines in the current childhood and adolescent immunization schedule overloads the immune system. Every day, a healthy baby's immune system successfully fights off thousands of antigens. Young infants build immunity to an estimated 100,000 different organisms. Even though the childhood and adolescent immunization schedule has over 30 injections, the 14 vaccines recommended for children represent only about 0.01% of immunity infants may develop. All 14 vaccines we give children now contain less than 150 antigens combined. As a comparison, one smallpox vaccine contained over 200 antigens alone. Vaccinations do not overwork the immune system!

allowing for easier access. A new way of vaccinating with the bifurcated needle was adopted. The bifurcated needle was more efficient and cost effective than the jet injector that had been used. And an important epidemiological tool, case surveillance, accompanied by mass vaccination started to turn the tide. The last naturally acquired case of smallpox occurred in 1975 and the last smallpox death occurred in 1977. The WHO declared the world free of smallpox on May 8, 1980. Although other eradication programs have been initiated (WHO-UNICEF campaign to eliminate yaws in 1952, WHO Global Malaria Eradication Programme in 1955, the Carter Center Guinea Worm Eradication Program in 1986, WHO Global Polio Eradication Initiative in 1988), no other human disease has ever been eradicated on a global scale.

## 2009 H1N1 SWINE FLU VACCINATION

The 2009 H1N1Swine flu pandemic was caused by a novel influenza A virus that was first detected in the U.S. in April of 2009. The virus contained unique genes not seen in humans or animals before, so there was no natural immunity to the disease. And because of the novelty of the virus, previous flu vaccines did not offer any cross-protection to the new virus. So, the disease spread much more quickly than a typical seasonal flu. Also, unlike seasonal influenza epidemics where most deaths occur in people 65 years and older, this virus pattern had higher mortality in people under 65, hitting primarily children, and young and middle-aged adults (much more similar to the 1918 influenza epidemic). Parts of the pan-flu plan were implemented, including closing schools for five days once outbreaks had started within them, to stop spread.

In August 2010, the WHO declared the 2009 influenza pandemic over, meaning this pandemic only lasted 16 months. The quick resolution of the pandemic can be attributed to the quick creation of an effective vaccine that elicited an excellent immune response, preventing disease spread. Three months into the pandemic, vaccines were in FDA clinical trials. The FDA approved four vaccines against the H1N1 Swine flu virus in September. The first doses of vaccine were given in the U.S. in early October. The number of cases started declining when vaccinations were widely available. Since the pandemic ended, this same virus strain has been circulating as part of seasonal flu. The 2009 H1N1 Swine flu pandemic vaccine (H1N1pdm09) has been included in seasonal flu vaccine every year. It is the A/Victoria/2570/2019 (H1N1)pdm09 strain in the 2022-2023 flu vaccine mentioned previously.

## COVID-19 VACCINATION

Remember from Chapter 5 that mRNA is a copy of the instructions of a gene used by ribosomes to build a protein? If not, how about a little refresher! The SARS-CoV-2 virus is an RNA virus, meaning the genetic material to make copies of itself is in RNA not DNA like a human. For replication, an RNA virus relies on our cells to crank out production of viral proteins instead of human proteins. This is done by the RNA from the virus entering the cell and hijacking the cell's machinery. From there, our cells produce all the viral proteins, which are assembled into new viruses, before escaping the infected cell. A smart move for a pathogen.

Learning from the RNA viruses, scientists set out to create vaccines and treatments using the same concepts. The idea was to use gene segments of mRNA from infectious pathogens injected into the host and have our cells manufacture the proteins and spit them back out of the cell. If the proteins were not human proteins, the immune system would recognize them as foreign and create an immune response to eliminate them. Let's say those proteins were from a virus. Our cells would make the viral proteins, release the proteins into the blood stream, and our immune system would identify them as foreign and develop antibodies to those proteins. The next time that protein was identified by the antibody, our immune system would mount a response to eliminate or lessen the effects of the virus. COVID-19 vaccines were made using mRNA technology.

In 1992, mRNA vaccine research was focused on creating vaccines for seasonal flu. This research never succeeded past rodent models. In 1995, mRNA was tested as a cancer treatment, also only in rodents. The problem this genius idea was that in real life when the mRNA from the vaccine or treatment was injected into a host, it was quickly degraded by the host cells. So, yes, a brilliant idea to monopolize on what viruses do and use mRNA to trick our cells to produce our own disease preventions and treatments, but it took a while to get the delivery right.

In 2012, the U.S. government started investing in the study of RNA vaccines and treatments. Around this time, a breakthrough in mRNA vaccine development technology occurred. There was a way to deliver the mRNA to the host without mRNA degradation: modify the mRNA or put it in lipid nanoparticle packages so the cells won't degrade them. **Advances in drug delivery systems have expedited the preclinical development of mRNA therapeutics.**

It wasn't until 2013 that an mRNA vaccine for prevention of an infectious disease, rabies, was tested in a human clinical trial. Unfortunately, this vaccine did not give participants the long-lasting immunity that was desired for such a fatal disease. During the outbreaks of SARS and Zika virus, mRNA vaccines were options to prevent the spread of disease, but it wasn't until the COVID-19 pandemic that dreams became reality. The thirty years of mRNA vaccine technology research was rapidly adapted for the COVID-19 pandemic. Although not new to research, the COVID-19 vaccines were the first mass produced mRNA vaccines for humans.

Within days of the completed sequence of the SARS-CoV-2 viral RNA becoming available online, many companies were already working on creating mRNA vaccines. Biotech companies Pfizer, BioNTech, and Moderna had created prototype vaccines within days of the virus's genome sequence becoming available online. Moderna collaborated with the US National Institute of Allergy and Infectious Diseases (NIAID) to conduct mouse studies and launch human trials, all within less than ten weeks. Pfizer-BioNTech was in human clinical trials of the mRNA vaccine and received emergency approvals within eight months.

The first mRNA vaccines for COVID-19 in the U.S. were available for distribution on December 12, 2020, under FDA's Emergency Use Authorization (EUA). The EUA was lifted for the first vaccine from Pfizer-BioNTech on August 23, 2021. This is the fastest a new vaccine has been created, tested, and approved in the U.S., all thanks to Operation Warp Speed. The process from FDA Phase I trial to approval normally takes over eight years. Two mRNA COVID-19 vaccines produced by Pfizer-BioNTech and Moderna had received full FDA approval by January 31, 2022. Between December 2020 and November 2022 an estimated 3.2 million deaths, over 18 million hospitalizations, and 119 million infections were averted due to the vaccinations given in the U.S. Thanks, Public Health!

In conclusion, after living through a pandemic where social media rants, dissemination of misinformation, and information being flat out withheld was the norm, I offer one final piece of advice. Please do your due diligence while reading information about any infectious disease outbreak and consider the sources (AND FUNDING SOURCES) of the information. Then you decide if the information is reliable.

**FIGURE 10.11** Comparison of Reported COVID-19 Incidence per 100,000 People in the United States and the Simulated Scenario without Vaccination.

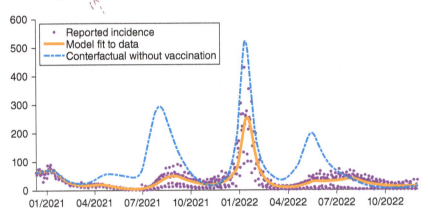

**MOVIE SPOTLIGHT:** *Hiden* (Duffer Brothers, 2015)
Causative agent – unknown, likely airborne
A family of three takes refuge in an abandoned fallout shelter shortly after the catastrophe to hide from the "Breathers" outside.
"One, never be loud." "Two, never lose control. Three, never open the door. Four, never talk about the Breathers."

**TV SERIES SPOTLIGHT:** *The Rest of Us* (Mazin & Druckman, 2023)
Causative agent – Cordyceps fungal infection
Twenty years into a pandemic a smuggler escorts a teenager with immunity from a quarantine zone across the country.
"I used to hate the world and I was happy when everyone died. But I was wrong. Because there was one person worth saving. That's what I did. I saved him. Then I protected him. That's why men like you and me are here. We have a job to do…"
The TV series is loosely based on the video game *The Rest of Us* (2013).

# CHAPTER 11

*With contributions from Travis Parsons, SSgt USAF former active duty, Aircrew Life Support Journeyman Graduate of Combat, Arctic, Water Survival Schools and certified Air Force Instructor*

## GET A KIT, MAKE A PLAN, BE PREPARED

### (OR…"I CAN'T COME IN TO WORK TODAY. I'M BUILDING MY ZOMBIE-PROOF BUNKER.")

"If you are generally well equipped to deal with a Zombie apocalypse you will be prepared for a hurricane, pandemic, earthquake, or terrorist attack."

– Ali S. Khan (RET), MD, MPH U.S. Assistant Surgeon General, Director, Office of Public Health Preparedness and Response

Courtesy of Ego

It is very simple in principle and takes a little discipline to put into practice. Surviving a zombie outbreak could be associated with the same principles that apply to any mass crisis situation. No matter if you are dealing with a hurricane, earthquake, or flesh-eating walking corpses, the first 72 hours of any breakdown in what we call our social norms will be the most critical, as it is during this time when you find initial cover, protection, and make a self-assessment of the situation. The difference between a combat situation and a zombie crisis is that you have no rescuing party coming in order to safely extract you from the impending danger. Danger will be all around you, which is exactly why applying combat survival situations will be much more critical.

 **FIGURE 11.1.** Your tax dollars at work.

## CDC EMERGENCY PREPAREDNESS PROCEDURES

### GET A KIT

If the proverbial fecal matter ever makes physical contact with the oscillating rotor blade, what will you do for food and water? What happens when the convenience of a full grocery store down the road suddenly doesn't exist? According to the CDC, 50% of U.S. adults aren't prepared for a real emergency, whether it's fire, flood, tornado, or undead cannibals.

Source: CDC

Contributed by Travis Parsons. © Kendall Hunt Publishing Company.

221

 FIGURE 11.2.

"I *assume* you remembered to pack the emergency squeaky bone?"

Source: Christopher Green

Ask yourself, if I had to bug out of everyday life in a moment's notice, with no warning, what would I grab? How far would I get? Where do I need to go? These few simple questions can be answered in advance, and your survival odds increased just by having a small amount of preparation in place. First, do you have a survival bag with which you could simply pick up and go? Think about what should be in that bag. What do you need to get by for a short time and seek out shelter for a temporary period?

The first step is to put together your emergency kit. This needs to be a comprehensive kit for any emergency, and needs to function if you need to shelter-in-place or quickly evacuate (the bug-out bag). The CDC offers a list of items it suggests for emergencies, but reminds you, rightly so, that each family is unique. Yours might need provisions not included on the list. Consider all emergencies, not just the moaning, shambling, biting kind. Finally, when preparing your kit, make sure you think about *all* your family members.

## MAKE A PLAN

So now you've got your trusty emergency zombie (and I suppose earthquake, etc.) kit, so now what? Time to plan for what to do when the dead decide to no longer stay dead.

Full disclosure: The authors were born in the dark ages before cell phones. It wasn't uncommon to know off the top of your head dozens if not hundreds of phone numbers. How about now? Ask yourself how many phone numbers you know. Bet you can count them on your hands and have some fingers left over.

**TABLE 11.1**                                                    Recommended Emergency Kit

| Water (One gallon per person, per day)<br>Food (2 weeks supply)<br>Tools and Supplies | Sanitation and Hygiene | Miscellaneous |
|---|---|---|
| Utility Knife | N95 or surgical masks | Whistle |
| Duct tape | Towels | Entertainment items |
| Plastic sheeting | Soap | Pet Supplies |
| Battery powered radio | Bleach | Extra house and car keys |
| Matches | First aid kit | Two way radios |
| Multipurpose tool | Medication (OTC and prescription) | |
| Flashlight | Clothing and Bedding | |
| Extra Batteries | Extra clothing, hat and sturdy shoes, rain gear | |
| Cell Phone w/ charger | Blankets or sleeping bags | |
| Maps | Tent | |
| Can opener | Important Documents | |
| Pot[1] | Copies of driver's license, birth certificate | |
| | Extra cash | |

[1] That's *a* pot. I think.

**FIGURE 11.3**                              Ignore warnings at your own peril.

So what happens if cell networks are down and you needed to contact someone via a landline? The CDC recommends first making a plan for you and your loved ones to connect with each other in case of an emergency. The next page shows the family emergency plan contact card, containing valuable information you may find yourself needing if you're on the business end of a horde of zombies. Pick a meeting place for your family, both close and farther away from your home. This would be a place where your entire family would know to meet during an emergency even if communications are down.

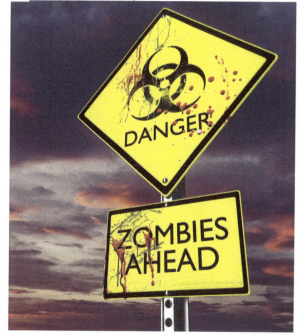

© Digital Storm/Shutterstock.com

FIGURE 11.4.

CDC family emergency plan contact card

# Ready
**Prepare. Plan. Stay Informed. ®**

# Family Emergency Plan

FEMA

Make sure your family has a plan in case of an emergency. Before an emergency happens, sit down together and decide how you will get in contact with each other, where you will go and what you will do in an emergency. Keep a copy of this plan in your emergency supply kit or another safe place where you can access it in the event of a disaster.

Out-of-Town Contact Name: _____    Telephone Number: _____

Email: _____

Neighborhood Meeting Place: _____    Telephone Number: _____

Regional Meeting Place: _____    Telephone Number: _____

Evacuation Location: _____    Telephone Number: _____

Fill out the following information for each family member and keep it up to date.

Name: _____    Social Security Number: _____
Date of Birth: _____    Important Medical Information: _____

Name: _____    Social Security Number: _____
Date of Birth: _____    Important Medical Information: _____

Name: _____    Social Security Number: _____
Date of Birth: _____    Important Medical Information: _____

Name: _____    Social Security Number: _____
Date of Birth: _____    Important Medical Information: _____

Name: _____    Social Security Number: _____
Date of Birth: _____    Important Medical Information: _____

Name: _____    Social Security Number: _____
Date of Birth: _____    Important Medical Information: _____

Write down where your family spends the most time: work, school and other places you frequent. Schools, daycare providers, workplaces and apartment buildings should all have site-specific emergency plans that you and your family need to know about.

**Work Location One**
Address: _____
Phone Number: _____
Evacuation Location: _____

**School Location One**
Address: _____
Phone Number: _____
Evacuation Location: _____

**Work Location Two**
Address: _____
Phone Number: _____
Evacuation Location: _____

**School Location Two**
Address: _____
Phone Number: _____
Evacuation Location: _____

**Work Location Three**
Address: _____
Phone Number: _____
Evacuation Location: _____

**School Location Three**
Address: _____
Phone Number: _____
Evacuation Location: _____

**Other place you frequent**
Address: _____
Phone Number: _____
Evacuation Location: _____

**Other place you frequent**
Address: _____
Phone Number: _____
Evacuation Location: _____

| Important Information | Name | Telephone Number | Policy Number |
|---|---|---|---|
| Doctor(s): | | | |
| Other: | | | |
| Pharmacist: | | | |
| Medical Insurance: | | | |
| Homeowners/Rental Insurance: | | | |
| Veterinarian/Kennel (for pets): | | | |

Dial 911 for Emergencies

Source: CDC

continued...

~~HUMAN~~ ZOMBIE BIOLOGY

# Family Emergency Plan

Make sure your family has a plan in case of an emergency. Fill out these cards and give one to each member of your family to make sure they know who to call and where to meet in case of an emergency.

ADDITIONAL IMPORTANT PHONE NUMBERS & INFORMATION:

< FOLD HERE >

ADDITIONAL IMPORTANT PHONE NUMBERS & INFORMATION:

## Family Emergency Plan

EMERGENCY CONTACT NAME:
TELEPHONE:

OUT-OF-TOWN CONTACT NAME:
TELEPHONE:

NEIGHBORHOOD MEETING PLACE:
TELEPHONE:

OTHER IMPORTANT INFORMATION:

Ready ®

DIAL 911 FOR EMERGENCIES

## Family Emergency Plan

EMERGENCY CONTACT NAME:
TELEPHONE:

OUT-OF-TOWN CONTACT NAME:
TELEPHONE:

NEIGHBORHOOD MEETING PLACE:
TELEPHONE:

OTHER IMPORTANT INFORMATION:

Ready ®

DIAL 911 FOR EMERGENCIES

ADDITIONAL IMPORTANT PHONE NUMBERS & INFORMATION:

< FOLD HERE >

ADDITIONAL IMPORTANT PHONE NUMBERS & INFORMATION:

## Family Emergency Plan

EMERGENCY CONTACT NAME:
TELEPHONE:

OUT-OF-TOWN CONTACT NAME:
TELEPHONE:

NEIGHBORHOOD MEETING PLACE:
TELEPHONE:

OTHER IMPORTANT INFORMATION:

Ready ®

DIAL 911 FOR EMERGENCIES

## Family Emergency Plan

EMERGENCY CONTACT NAME:
TELEPHONE:

OUT-OF-TOWN CONTACT NAME:
TELEPHONE:

NEIGHBORHOOD MEETING PLACE:
TELEPHONE:

OTHER IMPORTANT INFORMATION:

Ready ®

DIAL 911 FOR EMERGENCIES

## THE T ZONE

As we now know, humans can survive damage to the cerebrum and cerebellum. While popular culture shows us that *any* head shot will take a zombie down, it is more likely the only guaranteed way to put down the undead for good is to aim for what law enforcement and military refer to as the *T-zone*.

The T zone is an imaginary "T" between a target's eyes, continuing down to the tip of the nose. The reasons for aiming at your zombie is two-fold. First, the human skull is capable of deflecting even larger caliber rounds. Bounce a bullet off a zombie's skull and all you've succeeded in doing is getting his attention. The bone thickness in the T zone is minimal.

More importantly, however, hitting the undead in this T zone ensures the brainstem is destroyed. Even the most basic functions of zombies at least need the brainstem. So one shot one kill? Only if you cross your T's.

**FIGURE 11.4.**          Practice makes perfect.

© armi1961/Shutterstock.com

Think about the key elements to escaping any mass crisis situation. It is the mass populated areas where the chaos will first engulf society. You must have available enough food and water to sustain you and your family for at least 72 hours. The first 72 hours are the most critical to escape this area. In a combat survival situation, lost aircrew avoid mass populated areas and seek protection within the wild. They are trained to handle these very scenarios and are given the equipment to survive in the elements for quite some time while awaiting extraction. The only difference is that in a zombie outbreak, there is no hope of extraction to safety, so a short-term evasion and survival scenario will turn into an everyday lifestyle. Additionally, unlike aircrew, most of us do not simply carry around bagged water, fire-starting materials, first aid equipment, a sidearm, knife, emergency blanket and flares in everyday life. These basic survival materials must be proactively prepared ahead of time and in place where they may be accessed quickly.

Where are you going to go? Think about this for a moment. You have your survival bag, your loved ones, and are ready to make a quick exit of your suburban life in order to avoid the mass hysteria. You have nowhere to go. When in doubt, head to a camp-ground. They are secluded enough from the populated areas, where at the very least you can find temporary reprieve from the flesh buffet that your neighbors—

FIGURE 11.5

Want to feel old? Ask an
8-year-old what this is.

© Digital Storm/Shutterstock.com

who chose to hide out in their boarded-up homes—are offering. But this means scouting your destination. Take advantage of the normal days we have to know these coveted locations… where can you go, and most importantly, how can you get there and how can you get out? What is meant by this is, you must know more than one route to your temporary safety zone and more than one way out of that area. In most cases, it would be best to know at least three possible routes in and out. Expect the main and secondary roads to be clogged with mass hysteria. Stick to back roads and secluded avenues as much as possible. The bottom line: get as far away from the crisis area as quickly as you can and be prepared to get out of your temporary safety area just as quickly. Knowing how to get out of where you are going is equally as critical as getting there. Safety will always be temporary. In a zombie outbreak, danger will find you. Your advanced preparation is critical to your survival.

## BE PREPARED

Emergency preparedness kit? Check. Well thought out escape route? Check. Now the only question is… when do I use all this stuff? How will I know when I need to evacuate and grab the "bug-out bag" or just "shelter-in-place"?

Fire or police department warning procedures could include:

- Emergency Alert System (EAS)
- Outdoor warning sirens or horns
- News media sources
- NOAA Weather Radio alerts

**FIGURE 11.6.**

Red Cross Apps (American Red Cross, 2016)

© fad82/Shutterstock.com

But for up-to-the-minute information at your fingertips the American Red Cross offers free apps that give information on first aid for you and your pets, local shelters (along with which ones are equipped for medical emergencies and pets), and 35 different severe weather and emergency warnings (flood, fire, tornado, hurricane, wildfire, etc).

These warnings, of course, are only helpful to those who heed them. Ignore police and fire warnings at your own peril. If they say shelter, shelter. If they say bug-out, bug-out. "The hell with them, I know what I'm doing" is the equivalent of throwing the pin and holding onto the grenade.

## MEDICAL SUPPLIES

Hopefully, you've been able to take away some valuable information from this book on human and zombie biology. Remember, in the case of a zombie apocalypse, you are dealing with a breakdown of society and civilization on par with natural disasters, but you are also dealing with a contagion. What makes zombies extra frightening is not just that they are walking dead cannibals with a single-minded mission to feast on you, but that they can convert *you* too into a shambling, dead eating machine that will happily feast on your own family's flesh.

In this book you've learned about the possibilities, even the less-than-feasible ones. Is it a microbe being transmitted by bites? That's a bit more manageable; you need to establish a safety perimeter and keep the zombies away. Is it a microbe being transmitted by aerosols, or by a vector like mosquitoes? That's trickier. In fact, it might be a good idea to include mosquito netting in your emergency supplies, to fend off a multitude of mosquito-borne pathogens.

What kinds of medical supplies should you take along? Bleach is included in the CDC list for a good reason; it can destroy both bacteria and viruses, small amounts can be added to contaminated water for drinking (in a pinch), and a couple of inches can be used in a waste bucket with a lid if you can't leave your shelter to answer the call of nature. If you're going to be living off the land for awhile, you might want to have a supply of drinking water purification tablets, but that will only last so long. Even better, learn to build a still in order to distill pure water (the Mythbusters built one just out of plastic and duct tape, which is a great survival skill to know). Soap is important also; many people really underestimate the contribution soap makes to our life expectancy. Soap can lift germs away from surfaces and our skin, reducing the chances of infection. But say you have access to a pharmacy, what should you bring? It's not just the zombie microbe itself you have to worry about; what about all the other microbes out there, waiting to take you out just as decisively, if not as grotesquely, as the zombies? Well, as of the time of this writing, antibiotics are still critical to our health and survival. But right now, this very minute, they are losing their effectiveness due to the spread of antibiotic resistance genes between bacteria species. Even we can't predict how long antibiotics will continue to work; right now scientists are fearful that we are about to return to an age where a simple infection can kill a patient. We know that anti-viral medicines we have are fairly limited in their usefulness, as they tend to be highly specific to the viruses they can fight, and have very limited time-frames in which they are effective. Pain medicines would be important, as it is likely in a survival situation you or a family member could be injured or require surgery. Basic first-aid materials to staunch blood flow, wrap wounds, and bandage cuts would be beneficial. Some basic surgical equipment, such as scalpels and sutures, would be useful for more severe injuries. Finally, don't disregard vaccines. Some risks with vaccines are real, as mentioned, like cross-reactivity and allergic reactions to ingredients such as eggs. Other risks are entirely fictional, like the purported association between vaccines and autism. The risks are outweighed by the dangers of the disease itself... and remember, when you're vaccinated you're not only protecting yourself, but also your loved ones and community as well.

Remember, more information equals more chance for survival, and we sincerely hope you do survive with us. But just don't use the information to save *everyone*, because… well… what fun is a zombie apocalypse without zombies? Because if we're being completely honest, aren't we all rooting for it to happen… just a little bit?

> *"Everyone, deep in their hearts, is waiting for the end of the world to come."*
> —Haruki Murakami, writer

## WILL TO SURVIVE

In a zombie outbreak, tragic loss will be around you. From the early days of seeing it on nightly news reports, to the live-time situation of seeing your neighbor eaten alive while you can do nothing but watch helplessly, you have to simply get over it! You at that moment are alive and that is not something that can be easily disregarded. The will to survive is without a doubt the most important aspect of survival. Aircrew are instructed this, but instructing and applying are two separate beasts all together. If you allow your mind to get lost in the senseless and tragic events unfolding around you, then you too will be lost. When you find yourself in that sudden life and death situation where you must make a critical choice to survive, and you have chosen not to, then you have given up on survival and will succumb to the outbreak. Look ahead, want to survive, and consider each day a gift. Do not harp on the loss around you and instead be thankful that you have survived another day.

## THE ART OF THE KNIFE

You have a gun or two? That is great! When you run out of ammunition just go to any sporting goods store and pick up some more; the supply trucks will still be running… oh yeah. See the problem with relying on modern firearms? As society falls around us, we must adapt and change to landscape around us. Manufactured goods will be looted and over a short period of time, become obsolete. Firearms are good for the short-term route and good to have for that critical moment, but it is the knife that will get you by in the later days. A knife does not have to be reloaded. It will not jam on you and not suffer a mechanical failure. The knife will become your weapon of choice and be the most reliable component of your survival arsenal… (although a bullwhip could help). A knife will not give away your position and is something that someone of any fighting skill can quickly learn. Those who rely on modern technology will be the first survivors to fall. Learn medieval fighting tactics quickly and you will increase your ability to survive.

Remember what you have learned in this book about the brain of the zombie. They are using neuronal transmission, just like the rest of us. If they are truly dead, they are using some alternative to the energy molecule ATP, but they must be using neurotransmission to be directing movement and food-taxis signals in their old, reptilian brain areas, and then transmitting those signals to their muscles to generate the actual movement. If you can disrupt that transmission, you can stop a zombie. Severing the spinal cord, ideally with a beheading, can stop the transmission from the zombie brain to zombie muscles, and can stop the zombie. A chainsaw is only good until the gasoline runs out. A knife, or a sword, will be a tried-and-true zombie killer long into the zombie apocalypse.

# THIS IS NO TIME TO PANIC, PEOPLE!

We're coming up on the end of the book. But before everyone rushes out and forms neighborhood zombie patrols, the authors thought it best to reassure you that the zombie apocalypse is not actually imminent. We can summarize several scientific reasons you should keep your cool and just go about your daily lives without fretting about the dead returning.

## MUSCLES WILL NOT WORK WITHOUT THE CARDIOVASCULAR SYSTEM.

Just to be clear, the "Walking Dead" or "Night of the Living Dead" zombie scenario is absolutely implausible. Why? They always talk about shooting in the head but never mention the heart. As we said earlier, muscles don't work without ATP, and we don't make ATP if we don't breathe in oxygen, move that oxygen to our cells, and have food to use to generate ATP. The heart moves oxygen-rich blood to our cells, so our cells can make ATP. In someone with no working cardiovascular system, ATP generation, neurologic activity, and movement is just not possible. These shows never talk about that little hiccup.

## THE DEAD DON'T MAINTAIN HOMEOSTASIS.

We know it's unfair for us to expect you to remember things from earlier in the book, but jump back to chapter 1 and refresh your memory on homeostasis. Go ahead... we'll wait.

...

...

...

We back up to speed? Good. One of the fundamental characteristics of all living things is their ability to maintain stable internal levels for temperature, pH, etc. This maintenance is called homeostasis. But the dead? Nope. Once you shuffle off this mortal coil your body doesn't remain in homeostasis. So someone that is deceased becomes what the paramedics might call ART, or *At Room Temperature*.

So the dead don't maintain homeostasis. So what? Well, homeostasis is why you can survive outside in the winter. The undead? They'd be zombie popsicles at the first frost. And that's nothing compared to what happens when it gets warm. Bacteria excel at one thing. They eat, get twice as large as they should, and reproduce. They can accomplish this feat as rapidly as once every 20 minutes. It really wouldn't take long for soft tissues like muscles to begin to decay, to the point where the muscles turn mushy and can't contract any longer. Zombies lying around in a pile decaying a couple of hours after they are bitten: not that much of a catastrophe. You'd only have to avoid one for an hour or two.

## BITING IS INEFFICIENT.

So in popular zombie lore, a bite (or even a scratch) is always 100% effective at turning you into one of the undead. In reality, however, bites are an ineffective mode of transmission. You are unlikely, for instance, to contact rabies from an animal bite *even if that animal has rabies and is contagious*. So while the movies zombies spread like wildfire, it's implausible that a worldwide bloodborne pandemic spread by human bites where you can visually see the person is infected would spread so rapidly. Of course, HIV is an example of a worldwide bloodborne pathogen, but that took decades to reach pandemic proportions, and HIV is sneaky: the virus hangs out for years or decades causing no visible symptoms, and those infected are often unaware they have contracted the virus.

## IT'S A MINIMUM OF 24 HOURS FOR VIRAL INFECTION.

Something else Hollywood always gets wrong is the time it takes to infect. Times range from the implausible (6 hours in the *Dawn of the Dead* 2004 remake) to the absolutely ludicrous (10 seconds in 2013's *World War Z*). In reality viruses take significantly longer from infection to signs and symptoms. No virus on the planet works under 24 hours and most are significantly longer. Latency periods can range from weeks (Varicella zoster virus, causative agent of smallpox) to years (HIV, causative agent of AIDS). Prion diseases like Creutzfeldt-Jakob's disease can take up to 10 years before signs and symptoms develop.

However, before we start to rest on our laurels and put our guards down, we should let you know that...

## THIS IS THE PERFECT TIME TO PANIC, PEOPLE!

> FIGURE 11.7

Civil unrest, likely due to Black Friday sales.

Ververidis Vasilis / Shutterstock.com

## Zombie-like diseases aren't fictional... They exist today.

Rabies, toxoplasmosis, and kuru all are zombie-like diseases. While their present state is not capable of causing a pandemic, microorganisms mutate and evolve many times faster than we do. A little evolutionary push is all it would take to turn these diseases significantly more infectious, virulent, and deadly. Understanding how science works, knowing how your body gets its energy and how your immune system works; these are the biggest advantages you have in a pandemic today.

## Biting is inefficient, but there's always mutation.

A disease spread by bite isn't a real danger to the population. However, imagine these same diseases spread via droplet, fomite, fecal-oral, or worse, airborne transmission. Again, it's not unprecedented for evolutionary changes in transmission.

## We aren't always that good at managing the risks.

As mentioned, antibiotics have been a great boon to the health and life expectancy of us all. But we have used them inappropriately in our food chains, and forgotten that living things have resiliency. Our own carelessness with the use of antibiotics has accelerated the distribution of antibiotic resistance genes into pathogens, and despite warnings of the danger, we as a population, aren't doing very much to stop the acceleration. It might not be very long at all before we descend into a zombie apocalypse-like scenario, with pathogens we can no longer control affecting our society and those we love. The more people that understand the basic biology of humans, and zombies, the more people there are to help us reduce the risk of a return to pre-Industrial Revolution life expectancies.

## IN CONCLUSION...

Science is important, and it can be accessible to anyone. Rational thinking can save your life and the lives of those you love. Know where to get reliable information. Don't be swayed by agendas and avoid "click bait." Learn to distinguish useful information from biased statements. Carefully scrutinize graphs for source material and pay attention to the scale of the graphs. Don't assume you can take it at face value; a lot of people out there are trying to manipulate your emotions, rather than stimulate your intellect.

Pathogens are out there, and they can be dangerous. Right now, we tend to think we are able to control most bacterial infections, but that is a tenuous control. Beware of viruses, brain parasites, and especially prions. Each of these types of pathogens can be responsible for zombie-like diseases now, and can mutate in the future. Can radiation cause zombies? Implausible. Can radiation or natural selection alter already present pathogens into significantly deadlier strains? Absolutely. Understand how your immune system and normal flora defend you on a constant, daily basis. Also understand how we can protect infants, children, and the immunocompromised. Understand the benefits of vaccination and find out the real risks of the diseases they prevent. Know the difference between debunked, faked "science," and the real thing.

Get a kit, make a plan, be prepared. Don't wait until the start of the outbreak (or tornado, flash flood warning, etc.) to start scrambling for supplies. Know how to shelter in place, or where to go if you need to grab the "bug-out bag." Don't ignore emergency warnings. Don't put first-responders in danger because you wanted to wait it out. Remember, more information and scientific understanding means better chance for survival.

# CHAPTER 11

## QUESTIONS

1. According to the CDC, what percent of Americans are not prepared for a real emergency?

2. List ten essential things you should have in your emergency kit.

3. List three things you learned in this class which leads you to believe a zombie outbreak is not likely.

4. List three things you learned in this class which leads you to believe a zombie outbreak is at least plausible.

# CHAPTER 11

## PICK APART THE MOVIE!!! (GROUP PROJECT)

The winning movie from the chapter 4 worksheet will be screened in class. Take notes during the screening then get together with your group. Summarize the movie with the same criteria as the *Write a better zombie story* Worksheets. These criteria include:

- Classification of zombies as alive or dead
- Causative agent
- Transmission/epidemiology
- Treatment/immune response

Groups will pool their information into a cohesive synopsis of the scientific validity of the movie. This movie will be used as a template for future classes.

_____

_____

_____

_____

_____

_____

_____

_____

_____

## WRITE A BETTER ZOMBIE STORY PART IV
## (THE THRILLING CONCLUSION)

Now that you've written parts I and II, tie them together into a two to three page synopsis of your zombie book/movie/comic.

Of course we'll need an ending. Describe how to end the zombie pandemic. If it can't be stopped, you must explain how the usual treatments (vaccines, antibiotics, etc.) proved to be ineffective. Whether it can be stopped or not by science, give steps to prevent the spread of the infection. Please keep consistent with the previous sections.

Assume I'm a Hollywood producer. *Sell* it to me and make it believable. A catchy title wouldn't hurt your chances. Be sure to hit all the scientific topics from earlier parts including:

- Are your zombies truly dead? Dead-ish? Explain what you mean by that.
- How did the outbreak start?
- How is the disease transmitted?
- What is the treatment/cure if any?
- How can we survive?

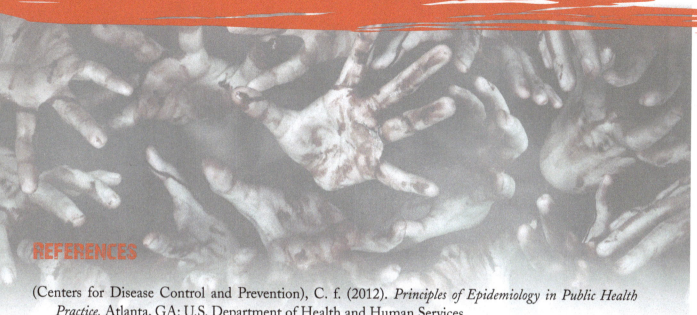

# REFERENCES

(Centers for Disease Control and Prevention), C. f. (2012). *Principles of Epidemiology in Public Health Practice.* Atlanta, GA: U.S. Department of Health and Human Services.

(Centers for Disease Control and Prevention), C. f. (2016, March 5). *Excellence in Curriculum Innovation through Teaching Epidemiology (EXCITE).* Retrieved from http://www.cdc.gov/EXCITE/epidemiology.html

Alberer, M., Gnad-Vogt, U., Hong, H. S., Mehr, K. T., Backert, L., Finak, G., Gottardo, R., Bica, M. A., Garofano, A., Koch, S. D., Fotin-Mleczek, M., Hoerr, I., Clemens, R., & von Sonnenburg, F. (2017). Safety and immunogenicity of a mRNA rabies vaccine in healthy adults: An open-label, non-randomised, prospective, first-in-human phase 1 clinical trial. *Lancet, 390,* 1511-1520.

American Lung Association. (2021). *Coronaviruses.* Retrieved from https://www.lung.org/lung-health-diseases/lung-disease-lookup/coronavirus

American Lung Association. (n.d.). *Preventing SARS.* Retrieved from http://www.lung.org/lung-health-and-diseases/lung-disease-lookup/severe-acute-respiratory-syndrome-sars/preventing-sars.html

American Red Cross. (2016, May 31). *WWW.Redcross.org.* Retrieved from http://www.redcross.org/get-help/prepare-for-emergencies/mobile-apps

Balaguero, J. (Director). (2007). *[Rec]* [Motion Picture].

Barford, E. (2013, September 18). Parasite makes mice lose fear of cats permanently. *Nature News.*

Bauman, R. W. (2017). *Microbiology with Diseases by Taxonomy* (5th ed). New York, NY: Pearson Education, Inc.

Belson, W. (1986). *Validity in survey research.* Brookfield, VT: Gower.

Benedetti, F. (2014). The serotonin transporter genotype modulates the relationship between. *Bipolar Disorders 16(8),* 857-866. doi:10.1111/bdi.12250

Boyle, D. (Director). (2002). *28 Days Later* [Motion Picture].

Brooks, M. (2003). *The Zombie Survival Guide: Complete Protection from the Living Dead.* New York: Crown Publishing.

Brooks, M. (2006). *World War Z: An Oral History of the Zombie War.* New York: Crown Publishing.

Bjornevik K, Cortese M, Healy BC, Kuhle J, Mina MJ, Leng Y, Elledge SJ, Niebuhr DW, Scher AI, Munger KL, Ascherio A. (2022) Longitudinal analysis reveals high prevalence of Epstein-Barr virus associated with multiple sclerosis. *Science.* Jan 21;375(6578):296–301. doi: 10.1126/science.abj8222. Epub 2022 Jan 13. PMID: 35025605.

Centers for Disease Control and Prevention. (2016, May 24). *2009 H1N1 Flu Vaccine*. Retrieved from CDC.gov: http://www.cdc.gov/h1n1flu/vaccination/

Centers for Disease Control and Prevention. (2016, May 24). *Emergency Preparedness and Response*. Retrieved from CDC.gov: http://emergency.cdc.gov/preparedness/index.asp

Centers for Disease Control and Prevention. (n.d.). *Glossary of Terms*. Retrieved March 15, 2016, from http://www.cdc.gov/hantavirus/resources/glossary.html

Centers for Disease Control and Prevention. (n.d.). *Healthcare-associated Infections (HAIs): Basic Infection Control and Prevention Plan for Outpatient Oncology Settings* . Retrieved March 15, 2016, from http://www.cdc.gov/HAI/settings/outpatient/basic-infection-control-prevention-plan-2011/transmission-based-precautions.html

Centers for Disease Control and Prevention. (2017). *Severe Acute Respiratory Syndrome (SARS)*. Retrieved from http://www.cdc.gov/sars/index.html

Centers for Disease Control and Prevention. (2019). *2009 H1N1 pandemic timeline*. Retrieved from https://www.cdc.gov/flu/pandemic-resources/2009-pandemic-timeline.html

Centers for Disease Control and Prevention. (2021). *History of smallpox*. Retrieved from https://www.cdc.gov/smallpox/history/history.html

Centers for Disease Control and Prevention. (2021). *Science Brief: SARS-CoV-2 and surface (fomite) transmission for indoor community environments*. Retrieved from https://www.cdc.gov/coronavirus/2019-ncov/more/science-and-research/surface-transmission.html

Centers for Disease Control and Prevention. (2022). *Disease impact of unsafe water*. Retrieved from https://www.cdc.gov/healthywater/global/disease-impact-of-unsafe-water.html

Centers for Disease Control and Prevention. (2023). *Influenza Type A Viruses*. Influenza (Flu). Retrieved from https://www.cdc.gov/flu/about/viruses/types.htm

Cook, T. L. (2015). A controlled prospective study of Toxoplasma gondii infection in individuals with schizophrenia: Beyond seroprevalence. *Journal of Psychiatric Research 60*, 87–94.

Craven, W. (Director). (1988). *The Serpent and the Rainbow (film)* [Motion Picture].

Crichton, M. (2002). *Prey*. New York: Harper Collins Publishers.

Davis, W. (1985). *The Serpent & the Rainbow*. New York: Simon & Schuster.

Dekker, F. (Director). (1986). *Night of the Creeps* [Motion Picture].

Dietter, S. (Director). (1996). *The Simpsons Season 7 Episode 23 - Much Apu About Nothing* [Motion Picture].

Druckman, N. (Creator). (2013). *The Last of Us* [Video Game]. Naughty Dog.

Dolgin, E. (2021). The tangled history of mRNA vaccines. *Nature, 597*, 318–324. doi: https://doi-org.ezp.lib.cwu.edu/10.1038/d41586-021-02483-w

Doucleff, M. (2014, October 2). No, Seriously, How Contagious is Ebola? NPR. Retrieved from http://www.npr.org/blogs/health/2014/10/02/352983774/no-seriously-how-contagious-is-ebola

Duffer, M. & Duffer, R. (Directors). (2015). *Hidden (film)* [Motion Picture].

Eberhardt, T. (Director). (1984). *Night of the Comet* [Motion Picture].

England, E. (Director). (2013). *Contracted* [Motion Picture].

Epidemiology, A. (2016, March 5). *Outbreaks, epidemics and pandemics—what you need to know*. Retrieved from http://www.apic.org/For-Consumers/Monthly-alerts-for-consumers/Article?id=outbreaks-epidemics-and-pandemicswhat-you-nee

Epomedicine. (2017, April 5). *Antigenic shift and drift* [Internet]. Retrieved from: https://epomedicine.com/medical-students/antigenic-shift-drift/

Faria, M. (013). Violence, mental illness, and the brain — a brief history of psychosurgery: Part 1— from trephination to lobotomy. *Surgical Neurology International 4(49)*. doi:10.4103/2152-7806.110146

Ferland, G. (Director). (2010). *The Walking Dead Season 1 Episode 6 - "TS-19"* [Motion Picture].

Fitzpatrick, M. C., Moghadas, S. M., Pandey, A., & Galvani, A. (2022). Two years of U.S. COVID-19 vaccines have prevented millions of hospitalizations and deaths. *Commonwealth Fund.* https://doi.org/10.26099/whsf-fp90

Flegr, J. (2002, July 2). Increased risk of traffic accidents in subjects with latent toxoplasmosis: A retrospective case-control study. *Biomed Central Infectious Disease*, 2-11.

Flegr, J. J. (2002). Increased risk of traffic accidents in subjects with latent toxoplasmosis: a retrospective case-control study. *BMC Infec. Dis.* 2(1), 11.

Fleischer, R. (Director). (2009). *Zombieland* [Motion Picture].

Food and Drug Administration. (2022). *Influenza vaccine for the 2022-2023 season.* Retrieved from https://www.fda.gov/vaccines-blood-biologics/lot-release/influenza-vaccine-2022-2023-season

Forbes. (2016, May 24). *forbes.com.* Retrieved from http://www.forbes.com/forbes/2010/0208/opinions-vaccine-flu-h1n1-health-on-my-mind.html

Forster, M. (Director). (2013). *World War Z* [Motion Picture].

Fothergill, A. (Director). (2006). *Planet Earth Video Series.* [Motion Picture].

Garden, C. (2014, November). A spirit of scientific rigor: Koch's postulates in twentieth-century medicine. *Microbes & Infection 16(11)*, 885-892. doi:10.1016/j.micinf.2014.08.012

Gardner, J. (1984). *Gilgamesh: Translated from the Sin-leqi-unninni version.* New York: Alfred A Knopf.

Garrett-Bakelman, F. E. (2019). The NASA Twins Study: A multidimensional analysis of a year-long human spaceflight. *Science 364.*

George, A. (2003). *The Epic of Gilgamesh.* New York: Penguin Books.

Gordon, S. (Director). (1985). *Re-Animator* [Motion Picture].

Green, C., Clark, K., Mathis, K., Barkhurst, S., & Mansfield, J. (2015). *Introductory Biology Laboratory Manual.* Dubuque, IA: Kendall Hunt.

Green, C. F. & Clark, K. (2016). *Microbiology for Health Professionals* (3rd ed rev). Dubuque, IA: Kendall Hunt.

Gregg, M. B. (2002). *Field Epidemiology* (2nd ed.). Oxford: Oxford University Press.

Guallar MP, Meiriño R, Donat-Vargas C, Corral O, Jouvé N, Soriano V. Inoculum at the time of SARS-CoV-2 exposure and risk of disease severity, Int J Infect Dis. 2020;97: 290–292. ISSN 1201-9712, https://doi.org/10.1016/j.ijid.2020.06.035.

Guay, D. (2008). Contemporary management of uncomplicated urinary tract infections. *Drugs 68*, 1169-1205.

Gunn, J. (Director). (2006). *Slither* [Motion Picture].

Halperin, V. (Director). (1932). *White Zombie* [Motion Picture].

Healthline. (2019, May 8). *Cotard delusion and walking corpse syndrome.* Retrieved from https://www.healthline.com/health/cotard-delusion

Helen C. Leggett, C. K. (2012). Mechanisms of Pathogenesis, Infective Dose and Virulence in Human Parasites. *PLoS Pathog*, e1002512.

Hinze-Selch, D. W. (2007). A Controlled Prospective Study of Toxoplasma gondii Infection in Individuals With Schizophrenia: Beyond Seroprevalence. *Schizophrenia Bulletin 33*(3), 782-788.

Horton, R. (2004). A statement by the editors of The Lancet. *The Lancet 363*(9411), 747-749.

Ingram, W. (2013, September 18). Mice infected with low-virulence strains of Toxoplasma gondii lose their innate aversion to cat urine, even after extensive parasite clearance. *PLoS One*, e75246. doi:10.1371/journal.pone.0075246.

Kandel, E., Schwartz, J., Jessell, T. (1991) *Principles of Neural Science.* 3rd edition. Elsevier.

Kao, C. (1986). Tetrodotoxin and the Haitian zombie. *Toxicon 24*, 747-749.

Kao, C. (1990). Tetrodotoxin in 'zombie powder'. *Toxicon 28*, 29-132.

Kean, S. (2014, May 6). Neuroscience's Most Famous Patient. *Slate Magazine*. Retrieved from http://www.slate.com/articles/health_and_science/science/2014/05/phineas_gage_neuroscience_case_true_story_of_famous_frontal_lobe_patient.html

Kilbourne, E. D. (2006). Influenza pandemics of the 20th century. *Emerging Infectious Diseases, 12*(1), 9–14. doi: 10.3201/eid1201.051254

Kirkman, R. (2003-present). *The Walking Dead (comic)*. Berkeley: Image Comics.

Knowledge, H. (2016, March 5). *Epidemic theory (effective and basic reproduction numbers, epidemic thresholds) and techniques for infectious disease data (construction and use of epidemic curves, generation numbers, exceptional reporting and identification of significant clusters)*. Retrieved from http://www.healthknowledge.org.uk/public-health-textbook/research-methods/1a-epidemiology/epidemic-theory

Korownyk, C. (2014). Televised medical talk shows— what they recommend and the evidence to support their recommendations: a prospective observational study. *British Medical Journal*, 349, g7346.

Labor, U. D. (2016, March 5). *OSHA Safety and Health Management Systems eTool*. Retrieved from https://www.osha.gov/SLTC/etools/safetyhealth/comp3.html

Lawrence, F. (Director). (2007). *I Am Legend* [Motion Picture].

Lazarevic V, Mantas I, Flais I, Svenningsson P. Fluoxetine Suppresses Glutamate- and GABA-Mediated Neurotransmission by Altering SNARE Complex. (2019). *International Journal of Molecular Sciences*. Aug 30;20(17):4247. doi: 10.3390/ijms20174247. PMID: 31480244; PMCID: PMC6747167.

Lucas, G. (Director). (1977). *Star Wars* [Motion Picture].

Levine AJ, Oren M. (2009). The first 30 years of p53: growing ever more complex. *Nature Reviews Cancer*. Oct;9(10):749-58. doi: 10.1038/nrc2723. PMID: 19776744; PMCID: PMC2771725.

María Pilar Guallar, Rosa Meiriño, Carolina Donat-Vargas, Octavio Corral, Nicolás Jouvé, Vicente Soriano. (2020). Inoculum at the time of SARS-CoV-2 exposure and risk of disease severity, *International Journal of Infectious Diseases*, 97(290-292) https://doi.org/10.1016/j.ijid.2020.06.035.

Mazin, C. & Druckman, N. (Directors). (2023). *The Last of Us* [TV Series].

McCullers, J. A. (2016). The role of punctuated evolution in the pathogenicity of influenza viruses. *Microbiology Spectrum, 4*(2). doi: 10.1128/microbiolspec.EI10-0001-2015

McDonald, B. (Director). (2008). *Pontypool* [Motion Picture].

Merriam-Webster. (2012). *Merriam-Webster dictonary*. Retrieved from Merriam-Webster.com

Mikami, S. (1996). Resident Evil (video game). Capcom.

Monikaben Padariya, Mia-Lyn Jooste, Ted Hupp, Robin Fåhraeus, Borek Vojtesek, Fritz Vollrath, Umesh Kalathiya, Konstantinos Karakostis. (2022). The Elephant Evolved p53 Isoforms that Escape MDM2-Mediated Repression and Cancer, *Molecular Biology and Evolution*, Volume 39, Issue 7, https://doi.org/10.1093/molbev/msac149

Munz, P. I. (2009). When zombies attack!: Mathematical modelling of an outbreak of zombie infection. *Infectious Disease Modelling Research Progrress*, 133-150.

Nelson, M. A. (2010). Brief communication: Mass spectroscopic characterization of tetracycline in the skeletal remains of an ancient population from Sudanese Nubia 350–550 CE. *American Journal of Physical Anthropology*, 151-154.

New York Police Department. (2016, March 20). *Felony crime statistics 2000-2014*. Retrieved from NYC.gov: http://www.nyc.gov/html/nypd/downloads/pdf/analysis_and_planning/seven_major_felony_offenses_2000_2014.pdf

Nolan, C. (Director). (2000). *Memento* [Motion Picture].

Nolan, C. (Director). (2008). *The Dark Knight* [Motion Picture].

O'Bannon, D. (Director). (1985). *Return of the Living Dead* [Motion Picture].

Organization of American States. (2013, 03 15). *Haitian Penal Code*. Retrieved from http://www.oas.org/juridico/mla/fr/hti/fr_hti_penal.html

Ottowa, U. (2016, March 5). *Society, the Individual, and Medicine - Epidemic Curves*. Retrieved from http://www.med.uottawa.ca/sim/data/Public_Health_Epidemic_Curves_e.htm

Padariya M, Jooste M.-L., Hupp T, Fåhraeus R, Vojtesek B, Vollrath F, Kalathiya U, Karakostis K. The elephant evolved p53 isoforms that escape MDM2-mediated repression and cancer. Mole Biol Evol. 2022 July;39(7): msac149, https://doi.org/10.1093/molbev/msac149

Peiris, J. S., Yuen, K. Y., Osterhaus, A. D., & Stöhr, K. (2003). The severe acute respiratory syndrome. *The New England Journal of Medicine, 349*(25), 2431-2441.

PREVAIL II Writing Group; Multi-National PREVAIL II Study Team, Davey RT Jr, Dodd L, Proschan MA, Neaton J, Neuhaus Nordwall J, Koopmeiners JS, Beigel J, Tierney J, Lane HC, Fauci AS, Massaquoi MBF, Sahr F, Malvy D. A (2016). Randomized, Controlled Trial of ZMapp for Ebola Virus Infection. *New England Journal of Medicine*. Oct 13;375(15):1448–1456. doi: 10.1056/NEJMoa1604330. PMID: 27732819; PMCID: PMC5086427.

Poirotte, C. P. (2016). Morbid attraction to leopard urine in Toxoplasma-infected chimpanzees. *Current Biology 26*(3), R98-R99.

Progress, A. (2016, March 1). *https://www.amprogress.org*. Retrieved from https://www.amprogress.org/animal-research-benefits

Public Health Agency of Canada. (n.d.). *CAMPYLOBACTER JEJUNI PATHOGEN SAFETY DATA SHEET - INFECTIOUS SUBSTANCES*. Retrieved March 15, 2016, from http://www.phac-aspc.gc.ca/lab-bio/res/psds-ftss/campylobacter-jejuni-eng.php

Public Health Agency of Canada. (n.d.). *EBOLAVIRUS PATHOGEN SAFETY DATA SHEET - INFECTIOUS SUBSTANCES*. Retrieved March 15, 2016, from http://www.phac-aspc.gc.ca/lab-bio/res/psds-ftss/ebola-eng.php

Ragona, U. (Director). (1964). *The Last Man on Earth* [Motion Picture].

Robinson WH, Steinman L. (2022) Epstein-Barr virus and multiple sclerosis. *Science*. Jan 21;375(6578): 264-265. doi: 10.1126/science.abm7930. Epub 2022 Jan 13. PMID: 35025606.

Romero, G. (Director). (1968). *Night of the Living Dead* [Motion Picture].

Romero, G. A. (Director). (1973). *The Crazies* [Motion Picture].

Rubin, J. (Director). (2014). *Zombeavers* [Motion Picture].

Saleh, A., Qamar, S., Tekin, A., Singh, R., & Kashyap, R. (2021) Vaccine development throughout history. *Cureus, 13*(7), e16635. DOI 10.7759/cureus.16635

Schmid-Hempel P, Frank SA. (2007) Pathogenesis, virulence, and infective dose. *PLoS Pathogens*. Oct 26;3(10):1372-3. doi: 10.1371/journal.ppat.0030147. PMID: 17967057; PMCID: PMC2042013.

Seabrook, W. (1929). *The Magic Island*. New York: Literary Guild of America; First Edition edition (1929).

Serretti, A. (2005). The influence of serotonin transporter promoter polymorphism (SERTPR) and other polymorphisms of the serotonin pathway on the efficacy of antidepressant treatments. *Progress in Neuropsychopharmacology & Biological Psychiatry*, 1074–1084.

Shaw, J. E. (2009 Mar 22). Parasite manipulation of brain monoamines in California killfish (*Fundulus parvipinnis*) by the trematode *Euhaplorchis californiensis*. *Proc Biol Sci*, 276(1659): 1137-1146.

Shaw, J. E. (2012). Brain-encysting trematodes and altered monoamine activity in naturally infected killifish Fundulus parvipinnis. *J Fish Biol, 81*(7), 2213–2222.

Shelley, M. W. (1818). *Frankenstein or The Modern Prometheus.* London: Lackington, Hughes, Harding, Mavor & Jones.

Snyder, Z. (Director). (2004). *Dawn of the Dead* [Motion Picture].

Sony Computer Entertainment America. (2014). The Last of Us - Remastered (video game).

Taubenberger, J. (2006). The origin and virulence of the 1918 Spanish Influenza Virus. *Proceedings of the American Philosophical Society 150*(1), 86-112.

Taubenberger, J. (2006). 1918 influenza: The mother of all pandemics. *Emerging Infectious Diseases*, 12, 15-22.

Thorn, C. R., Sharma, D., Combs, R., Bhujbal, S., Romine, J., Zheng, X., Sunasara, K., & Badkar, A. (2002). The journey of a lifetime - development of Pfizer's COVID-19 vaccine. *Current Opinion in Biotechnology, 78*(102803). doi: 10.1016/j.copbio.2022.102803.

Vaughn, M. (Director). (2014). *Kingsman: The Secret Service* [Motion Picture].

Wakefield, A. e. (1998). RETRACTED: Ileal-lymphoid-nodular hyperplasia, non-specific colitis, and pervasive developmental disorder in children. *The Lancet 351*(9103), 637-641.

Whedon, J. (Director). (2005). *Serenity* [Motion Picture].

WHO. (2017, July). *Antibiotic Resistance Fact Sheet.* Retrieved from http://www.who.int/en/news-room/fact-sheets/detail/antibiotic-resistance

Wright, E. (Director). (2004). *Shaun of the Dead* [Motion Picture].

Wood, E. D. (Director). (1959). *Plan 9 from Outer Space* [Motion Picture].

World Health Organization. (n.d.). *Timeline: WHO's COVID-19 response.* Retrieved from https://www.who.int/emergencies/diseases/novel-coronavirus-2019/interactive-timeline#event-50

Wu, F., Zhao, S., Yu, B., Chen, Y. M., Wang, W., Song, Z. G., Hu, Y., Tao, Z. W., Tian, J. H., Pei, Y. Y., Yuan, M. L., Zhang, Y. L., Dai, F. H., Liu, Y., Wang, Q. M., Zheng, J. J., Xu, L., Holmes, E. C., & Zhang, Y. Z. (2020). A new coronavirus associated with human respiratory disease in China. *Nature, 579*(7798), 265-269. doi: 10.1038/s41586-020-2008-3

Yeon, S.-h. (Director). (2016). *Train to Busan* [Motion Picture].

*Zombie Research Society.* (2016, March 18). Retrieved from Zombie Research Society homepage: http://zombieresearchsociety.com/advisory-board

# GLOSSARY

**Acetylcholine**—neurotransmitter that causes muscle action by transmitting nerve impulses across synapses.

**Acquired (specific) immunity**—immunity arising from exposure to antigens.

**Active immunity**—immunity in an organism resulting from its own production of antibody or lymphocytes.

**Agglutinate**—to clump or cause to clump, as bacteria or blood platelets.

**Anabolism**—constructive metabolism; the synthesis in living organisms of more complex substances from simpler ones (opposed to catabolism).

**Antibiotic**—any of a large group of chemical substances, as penicillin or streptomycin, produced by various microorganisms and fungi, having the capacity in dilute solutions to inhibit the growth of or to destroy bacteria and other microorganisms, used chiefly in the treatment of infectious diseases.

**Antibiotic resistance**—acquired ability of certain bacteria to withstand effects of antibiotics.

**Antibody**—any of numerous Y-shaped protein molecules produced by B cells as a primary immune defense, each molecule and its clones having a unique binding site that can combine with the complementary site of a foreign antigen, as on a virus or bacterium, thereby disabling the antigen and signaling other immune defenses.

**Antigen**—any substance that can stimulate the production of antibodies and combine specifically with them.

**Artificially acquired active immunity**—immunization with an antigen (vaccine).

**Artificially acquired passive immunity**—injection of antibody-containing serum, or immune globulin (IG), from another person or animal.

**Asexual reproduction**—reproduction, as budding, fission, or spore formation, not involving the union of gametes.

**ADP**—adenosine diphosphate, derived from ATP and serving to transfer molecular energy.

**Atom**—the smallest component of an element having the chemical properties of the element, consisting of a nucleus containing combinations of neutrons and protons and one or more electrons bound to the nucleus by electrical attraction; the number of protons determines the identity of the element.

**Atomic mass**—the average total number of neutrons and protons in the nucleus of a particular atom.

**Atomic number**—the number of positive charges or protons in the nucleus of an atom of a given element, and therefore also the number of electrons normally surrounding the nucleus.

**ATP**—adenosine triphosphate, serving as a source of energy for physiological reactions, especially muscle contraction.

**ATP synthase**—enzyme in the inner membranes of mitochondria and chloroplasts that creates the energy storage molecule adenosine triphosphate (ATP).

**Autism**—a pervasive developmental disorder of children, characterized by impaired communication, excessive rigidity, and emotional detachment: now considered one of the autism spectrum disorders.

**Axon**—the appendage of the neuron that transmits impulses away from the cell body.

**Axon terminal**—distal terminations of the branches of an axon.

**B lymphocyte**—developed in bone marrow that circulates in the blood and lymph and, upon encountering a particular foreign antigen differentiates into a clone of plasma cells that secrete a specific antibody and a clone of memory cells that make the antibody on subsequent encounters.

**Bacteriophage**—any of a group of viruses that specifically infect bacteria.

**Basophil**—leukocyte that causes inflammation.

**Behavior**—the aggregate of responses to internal and external stimuli.

**Bias**—a particular tendency, trend, inclination, feeling, or opinion, especially one that is preconceived or unreasoned.

**Binary fission**—asexual reproduction in unicellular organisms by division into two daughter cells.

**Bond**—a mutual attraction between two atoms resulting from a redistribution of their outer electrons.

**Capsid**—the coiled or polyhedral structure, composed of proteins that encloses the nucleic acid of a virus.

**Case**—a laboratory confirmed positive person.

**Catabolism**—destructive metabolism; the breaking down in living organisms of more complex substances into simpler ones, with the release of energy (opposed to anabolism).

**Cell wall**—the definite boundary or wall that is part of the outer structure of certain cells, as a plant cell.

**Cellular respiration**—the oxidation of organic compounds that occurs within cells, producing energy for cellular processes.

**Cerebellum**—a large portion of the brain, serving to coordinate voluntary movements, posture, and balance in humans, being in back of and below the cerebrum and consisting of two lateral lobes and a central lobe.

**Cerebrum**—the anterior and largest part of the brain, consisting of two halves or hemispheres and serving to control voluntary movements and coordinate mental actions.

**Chloroplast**—organelle in plant cells containing chlorophyll; site of photosynthesis.

**Containment strategies**—used to prevent transmission of infection, usually at the very beginning of an outbreak or epidemic.

**Complement**—a group of proteins found in normal blood serum and plasma that are activated sequentially in a cascadelike mechanism that allows them to combine with antibodies and destroy pathogenic bacteria and other foreign cells.

**Control**—A standard of comparison for checking or verifying the results of an experiment.

**Covalent bond**—the bond formed by the sharing of a pair of electrons by two atoms.

**COVID-19**—the name of the disease caused by the coronavirus SARS-CoV-2, and is short for Coronavirus Disease 2019.

**Cytoskeleton**—a shifting lattice arrangement of structural and contractile components distributed throughout the cell cytoplasm, composed of microtubules, microfilaments, and larger filaments, functioning as a structural support and transport mechanism.

**Cytosol**—the water-soluble components inside a cell, constituting the fluid portion that remains after removal of the organelles and other intracellular structures.

**Data**—individual facts, statistics, or items of information (plural of datum).

**Dendrite**—the branching process of a neuron that conducts impulses toward the cell.

**Deoxyribonucleic acid (DNA)**—deoxyribonucleic acid: an extremely long macromolecule that is the main component of chromosomes and is the material that transfers genetic characteristics in all life forms.

**Dependent variable**—(in an experiment) the event studied and expected to change when the independent variable is changed.

**Dermis**—the dense inner layer of skin beneath the epidermis, composed of connective tissue, blood and lymph vessels, sweat glands, hair follicles, and an elaborate sensory nerve network.

**Dopamine**—neurotransmitter acting within the brain to help regulate movement and emotion.

**Electron**—an elementary particle with a negative charge orbiting around the nucleus of an atom.

**Electron shell**—orbit taken by electrons around a nucleus.

**Elimination**—complete, intentional reduction of disease in a defined geographic area; action must still be taken to prevent spread of the disease.

**Emergency use authorization (EUA)**—an FDA status that allows for treatments/preventions to be used before receiving full approval

**Endoplasmic reticulum**—a network of tubular membranes within the cytoplasm of the cell, occurring either with a smooth surface (smooth endoplasmic reticulum) or studded with ribosomes (rough endoplasmic reticulum) involved in the transport of materials.

**Envelope**—the outer structure that encloses the nucleocapsids of some viruses.

**Enzyme**—any of various proteins, as pepsin, originating from living cells and capable of producing certain chemical changes in organic substances by catalytic action, as in digestion.

**Eosinophil**—leukocyte that can leave the bloodstream and phagocytize parasitic worms.

**Epidermis**—the outer, nonvascular, nonsensitive layer of the skin, covering the true skin or corium.

**Eradication**—complete, intentional, and permanent reduction of a disease worldwide; the disease-causing agent is no longer found.

**Erythrocyte**—red blood cell; enucleate disk concave on both sides, containing hemoglobin, and carrying oxygen to the cells and tissues and carbon dioxide back to the respiratory organs.

**Etiology**—study of the cause of disease.

**EUA**—Emergency Use Authorization

**Eukaryote**—any organism having as its fundamental structural unit a cell type that contains specialized organelles in the cytoplasm, a membrane-bound nucleus enclosing genetic material organized into chromosomes, and an elaborate system of division by mitosis or meiosis. Animals, plants, fungi, algae and protozoans.

**Evolution**—change in the gene pool of a population from generation to generation by such processes as mutation, natural selection, and genetic drift.

**Fever**—systemic increase in body temperature in response to pathogenic invasion.

**Flattening the curve**—slowing infection spread to reduce the number of daily cases and related demands on hospitals and infrastructure.

**Fomites**—objects which can become contaminated with infectious particles and transmit disease (tables, door handles, etc.), inanimate objects.

**Formed elements**—solid structures in plasma; erythrocytes, leukocytes, and platelets.

**Genetic drift**—minor genetic mutations that can accumulate and slowly change influenza viruses.

**Genetic shift**—major genetic mutations that can occur when one or more influenza viruses share genetic material.

**Golgi apparatus**—an organelle, consisting of layers of flattened sacs, that takes up and processes secretory and synthetic products from the endoplasmic reticulum and then either releases the finished products into various parts of the cell cytoplasm or secretes them to the outside of the cell.

**Graph**—a diagram representing a system of connections or interrelations among two or more things by a number of distinctive dots, lines, bars, etc.

**Helper T cell**—a T cell that stimulates B cells to produce antibody against a foreign substance.

**Herd (community) immunity**—the immunity or resistance to a particular infection that occurs in a group of people or animals when a very high percentage of individuals have been vaccinated or previously exposed to the infection.

**Hippocampus**—an enfolding of cerebral cortex into the lateral fissure of a cerebral hemisphere, having the shape in cross section of a sea horse. It is involved with the formation of new memories, along with learning and emotions.

**Histamine**—chemical released during allergic and inflammatory reactions, causing dilation of small blood vessels and smooth muscle contraction.

**Homeostasis**—the tendency of a system, especially the physiological system of higher animals, to maintain internal stability, owing to the coordinated response of its parts to any situation or stimulus that would tend to disturb its normal condition or function.

**Hydrophobic**—molecule having little or no affinity for water.

**Hypothesis**—testable proposition set forth as an explanation for the occurrence of some specified group of phenomena.

**Immunocompromised**—having an impaired or compromised immune response; immunodeficient.

**Incubation period**—the time between exposure and onset of symptoms.

**Independent variable**—(in an experiment) a variable that is intentionally changed to observe its effect on the dependent variable.

**Infection control measures**—hand washing, social distancing, etc.

**Inflammation**—a localized increase in temperature which helps repair damaged tissue.

**Innate (nonspecific) immunity**—immunity that occurs naturally as a result of a person's genetic constitution or physiology and does not arise from a previous infection or vaccination.

**Inorganic**—noting or pertaining to compounds that are not hydrocarbons or their derivatives.

**Interferon**—any of various proteins, produced by virus-infected cells that inhibit reproduction of the invading virus and induce resistance to further infection.

**Ion**—an electrically charged atom or group of atoms formed by the loss or gain of one or more electrons.

**Ionic bond**—the electrostatic bond between two ions formed through the transfer of one or more electrons.

**Isolation**—persons with laboratory-confirmed disease who have been directed to stay at home until they have recovered (10 days or so).

**Isotope**—any of two or more forms of a chemical element, having the same number of protons in the nucleus, or the same atomic number, but having different numbers of neutrons in the nucleus, or different atomic weights.

**Leukocyte**—any of various nearly colorless cells of the immune system that circulate mainly in the blood and lymph and participate in reactions to invading microorganisms or foreign particles.

**Lipid**—any of a group of organic compounds that are greasy to the touch, insoluble in water, and soluble in alcohol and ether.

**Lymphocyte**—a type of white blood cell having a large, spherical nucleus; T cells and B cells.

**Lytic cycle**—the viral reproductive cycle in which a virus takes over all metabolic activities of a cell; replicates many times and destroy its host cell.

**Lysosome**—a cell organelle containing enzymes that digest particles and that disintegrate the cell after its death.

**Macrophage**—a large white blood cell, occurring principally in connective tissue and in the bloodstream, that ingests foreign particles and infectious microorganisms by phagocytosis.

**Mean**—a quantity having a value intermediate between the values of other quantities; an average.

**Median**—the middle number in a given sequence of numbers, taken as the average of the two middle numbers when the sequence has an even number of numbers.

**Medulla oblongata**—the lowest or hindmost part of the brain, continuous with the spinal cord.

**Memory cell**—any small, long-lived lymphocyte that has previously encountered a given antigen and that on reexposure to the same antigen rapidly initiates the immune response (memory T cell) or proliferates and produces large amounts of specific antibody (memory B cell) the agent of lasting immunity.

**MERS**—the name of the disease caused by the coronavirus MERS-CoV, and is short for Middle East Respiratory Syndrome.

**Metabolism**—the sum of the physical and chemical processes in an organism by which its material substance is produced, maintained, and destroyed, and by which energy is made available.

**Midbrain**—the middle of the three primary divisions of the brain in the embryo of a vertebrate or the part of the adult brain derived from this tissue; mesencephalon.

**Mitigation strategies**—redirection of resources when assuming everyone in the community is infected.

**Mitochondrion**—an organelle in the cytoplasm of cells that functions in energy production; site of cellular respiration.

**Monocyte**—a large, circulating white blood cell, formed in bone marrow and in the spleen that ingests large foreign particles and cell debris.

**Mutation**—a sudden departure from the parent type in one or more heritable characteristics, caused by a change in a gene or a chromosome.

**Myelin sheath**—a wrapping of myelin around certain nerve axons, serving as an electrical insulator that speeds nerve impulses to muscles and other effectors.

**Natural killer (NK) cell**—a small killer cell that destroys virus-infected cells or tumor cells without activation by an immune system cell or antibody.

**Naturally acquired active immunity**—the natural exposure to an infectious agent or other antigen by the body.

**Natural selection**—the process by which forms of life having traits that better enable them to adapt to specific environmental pressures, as predators, changes in climate, or competition for food or mates, will tend to survive and reproduce in greater numbers than others of their kind, thus ensuring the perpetuation of those favorable traits in succeeding generations.

**Neurotransmitter**—any of several chemical substances, as epinephrine or acetylcholine, that transmit nerve impulses across a synapse to a postsynaptic element, as another nerve, muscle, or gland.

**Neutron**—an elementary particle having no charge; located in the nucleus along with protons.

**Neutrophil**—leukocyte that phagocytize invaders or kill by producing molecules of hydrogen peroxide and hypochlorite.

**Nonpolar molecule**—a molecule with no difference in local charge from one end to the other.

**Normal microbiota**—organisms that colonize the body's surfaces without normally causing disease.

**Novel**—a virus that has not previously affected humans (as far as current medical science can determine), a "new to human" virus subtype or strain

**Nucleotide**—any of a group of molecules that, when linked together, form the building blocks of DNA or RNA: composed of a phosphate group, the bases adenine, cytosine, guanine, and thymine, and a pentose sugar, in RNA the thymine base being replaced by uracil.

**Nucleus (atom)**—the positively charged mass within an atom, composed of neutrons and protons, and possessing most of the mass but occupying only a small fraction of the volume of the atom.

**Nucleus (cell)**—a specialized, usually spherical mass of protoplasm encased in a double membrane, and found in most living eukaryotic cells, directing their growth, metabolism, and reproduction.

**Operation Warp Speed**—a partnership between the Departments of Health and Human Services (HHS) and Defense (DOD) aimed to deliver 300 million doses of a safe, effective vaccine for COVID-19 by January 2021, as part of a broader strategy to accelerate the development, manufacturing, and distribution of COVID-19 vaccines, therapeutics, and diagnostics; initiated May 15, 2020

**Opportunistic pathogen**—non-pathogenic microorganism that acts as a pathogen in certain circumstances such as in an immunocompromised patient.

**Organelle**—a specialized part of a cell having some specific function; a cell organ.

**Organic**—noting or pertaining to a class of chemical compounds that formerly comprised only those existing in or derived from plants or animals, but that now includes all other compounds of carbon.

**Passive immunity**—immunity resulting from the injection of antibodies or sensitized lymphocytes from another organism or, in infants, from the transfer of antibodies through the placenta or from colostrum.

**Pathogen**—any disease-producing agent, especially a virus, bacterium, or other microorganism.

**PHEIC**—Public Health Emergency of International Concern.

**Penicillin**—any of several antibiotics of low toxicity, produced naturally by molds of the genus Penicillium.

**Personal protective equipment (PPE)**—gowns, face shields, masks, gloves

**Phagocytosis**—the ingestion of a smaller cell or cell fragment, a microorganism, or foreign particles by means of the local infolding of a cell's membrane and the protrusion of its cytoplasm around the fold until the material has been surrounded and engulfed by closure of the membrane and formation of a vacuole: characteristic of amoebas and some types of white blood cells.

**Photosynthesis**—the complex process by which carbon dioxide, water, and certain inorganic salts are converted into carbohydrates by green plants, algae, and certain bacteria, using energy from the sun and chlorophyll.

**Plasma (cell) membrane**—the selectively permeable membrane enclosing the cytoplasm of a cell.

**Platelet**—any of numerous, minute, protoplasmic bodies in mammalian blood that aid in coagulation.

**Polar molecule**—a molecule with a localized positive and negative end.

**Pons**—a band of nerve fibers in the brain connecting the lobes of the midbrain, medulla, and cerebrum.

**Population**—the assemblage of a specific type of organism living in a given area.

**Presymptomatic transmission**—transmission of infection before the onset of symptoms.

**Prokaryote**—any cellular organism that has no nuclear membrane, no organelles in the cytoplasm except ribosomes, and has its genetic material in the form of single continuous strands forming coils or loops. Bacteria and archaea.

**Proton**—an elementary particle with a positive charge in the nucleus of an atom along with neutrons.

**Qualitative data**—extremely varied in nature; includes virtually any information that can be captured that is not numerical in nature.

**Quantitative data**—anything that can be expressed as a number, or quantified.

**Quarantine**—separating and restricting the movement of people exposed (or potentially exposed) to a contagious disease; historically, quarantine is 2 weeks.

**$R_0$ (the reproductive rate or "R naught")**—the number of cases, on average, an infected person will cause during their infection; how many friends the disease will be shared with.

**Receptor**—an end organ or a group of end organs of sensory or afferent neurons, specialized to be sensitive to stimulating agents, as touch or heat.

**Recombination**—a shuffling of genetic information that happens when two different strains of a virus infect the same host cell, resulting in a new combination of genetic information in the daughter strain.

**Reservoir**—where infectious organisms are naturally found (air, water, animal), hosts may or may not get sick from the organism.

**Reuptake transporter**—responsible for reabsorption of a neurotransmitter at the axon terminal.

**Ribonucleic acid (RNA)**—any of a class of single-stranded molecules transcribed from DNA containing a linear sequence of nucleotide bases that is complementary to the DNA strand from which it is transcribed.

**Ribosome**—a tiny organelle occurring in great numbers and functioning as the site of protein manufacture.

**Rough endoplasmic reticulum**—*see endoplasmic reticulum.*

**SARS**—the name of the disease caused by the coronavirus SARS-CoV, and is short for Severe Acute Respiratory Syndrome.

**Schwann cell**—a cell of the peripheral nervous system that wraps around a nerve fiber, jelly-roll fashion, forming the myelin sheath.

**Serotonin**—a neurotransmitter that is involved in sleep, depression, memory, and other neurological processes.

**Sexual reproduction**—reproduction involving the union of gametes.

**Smooth endoplasmic reticulum**—*see endoplasmic reticulum.*

**Social distancing**—increasing distance between people in settings where people commonly come into close contact with one another (6 feet for droplet precautions).

**Soma**—the body of a neuron as contrasted with its appendages.

**Spikes (glycoproteins)**—appendages on the surface of viral envelopes containing a carbohydrate combined with a simple protein; used for attachment to a host.

**Standard deviation**—a measure of dispersion in a frequency distribution, equal to the square root of the mean of the squares of the deviations from the arithmetic mean of the distribution.

**Synaptic cleft**—the small gap, measured in nanometers, between an axon terminal and any of the cell membranes in the immediate vicinity.

**T lymphocyte**—lymphocytes developed in the thymus that circulate in the blood and lymph and orchestrate the immune system's response to infected or malignant cells.

**Toll-like receptor (TLR)**—numerous related proteins implicated in the development and defense of plants and animals.

**Transcription**—the process by which genetic information on a strand of DNA is used to synthesize a strand of complementary RNA.

**Translation**—the process by which a messenger RNA molecule specifies the linear sequence of amino acids on a ribosome for protein synthesis.

**Transmission**—how a disease spreads.

**Vaccine**—any preparation used as a preventive inoculation to confer immunity against a specific disease, usually employing an innocuous form of the disease agent, as killed or weakened bacteria or viruses, to stimulate antibody production.

**Vacuole**—a membrane-bound cavity within a cell, often containing a watery liquid or secretion.

**Valence shell**—outermost electron shell; contains electrons that can be transferred to or shared with another atom.

**Variable**—a quantity or function that may assume any given value or set of values.

**Virulence**—the severity of a pathogen.

**Zoonosis**—infectious disease that can be transmitted from animals to humans.

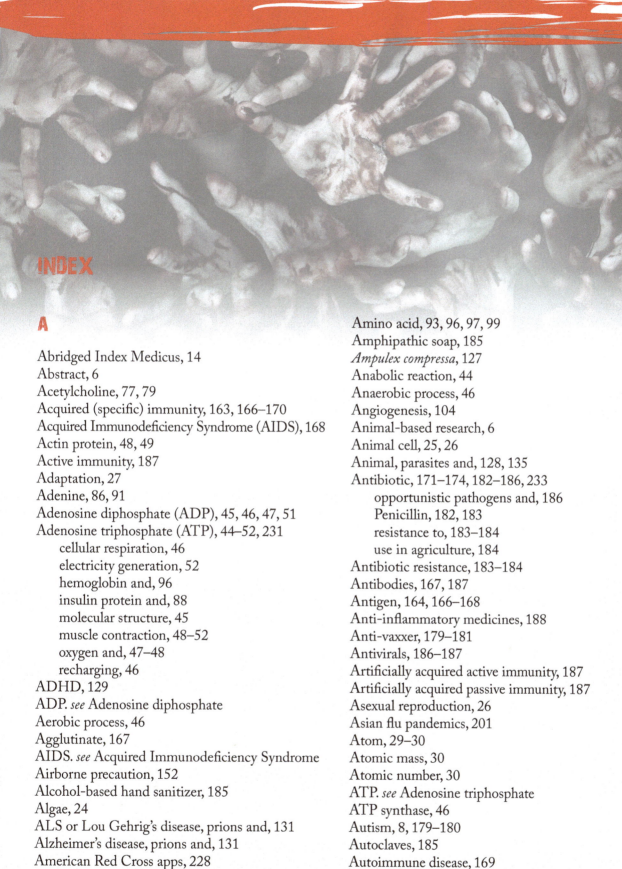

# INDEX

fomites, 197
genetic/antigenic drift, 198
H&N proteins, 198, 199
Hong Kong flu pandemics, 201
natural reservoirs, 197
pan flu plans, 201–203
Swine flu pandemics, 200
Information
finding reliable, 5
sources for reliable, 13–14
Infrared radiation, 102
Innate (nonspecific) immunity, 161–165
Inner mitochondrial membrane (IMM), 46, 47
Inorganic, 33
Integral protein, 26
Interferon, 163
International Space Station, 103
Introduction, 6
Ion, 31, 72
Ionic bond, 32
Isotope, 30

## K

Kelly, Mark, 103
Kelly, Scott, 103
Keratin, 161
Khan, Ali S., 220
King, Alexander, 64
*Kingsman: The Secret Service*, 154
Knife, for emergency survival, 230
Kuru, 132, 235

## L

*The Lancet*, 8, 179, 180
Language and communication, 66
*The Last Man on Earth*, 120
Learning, 66
Leprosy, 117–119
*Leucochloridium paradoxum*, 127
Leukocyte, 163, 164
Limbaugh, Rush, 200
Limbic system, 70
Line graph, 17
Lipid, 24, 36
Literature cited, 6
Living cell, 24

Living organism
characteristics of, 23–27
chemistry and, 27–38
Lobotomy, 66–67
Long-term memory, 68
Lymphocyte, 163, 166
Lysosome, 24, 25
Lytic cycle, 121
Lytic cycle of enveloped animal viruses, 122

## M

Macromolecules, 34
Macrophage, 163, 164
Mad cow disease, 129, 131
*Magic Island*, 76
Malaria, 134, 151
Materials and methods, 6
Mazin, Craig, 216
Mean, 17
Measles, mumps, and rubella (MMR) vaccine, 8
Measles virus, 134, 169, 178
Median, 17
Medulla oblongata, 68
Memory, 66, 68
Memory B cell, 168, 177
Memory cell, 168
Memory T cell, 168, 177
Meningitis, 164–165
Mental illness, 71
Metabolism
ATP and, 44–52
defined, 26
enzymes and, 43–44
*Methicillin Resistant Staphylococcus aureus* (MRSA), 117, 184
Microorganism, 114, 171
Microsoft Excel, 18
Micro-waves, 102
Midbrain, 68, 70
Mind-control, 135
Mitigation strategies, 210
Mitochondrion, 24
MMR, 177, 179
Mode of transmission, 147
Molecules, 34
Monocyte, 163, 164
*Moonlight Sonata*, 87, 88

Mosquito, 150–151
MRNA, 90, 91, 92–96
MRSA, 117, 184
Mucous membrane, 162
Multicellular organism, 24
Multiple sclerosis, 169
Mumps virus, 169
Murakami, Haruki, 232
Mutation, 86, 96–101
*Mycobacterium leprae*, 117
Myelin sheath, 73
Myosin, 48, 49

## N

NAD+, 46
NADH, 46, 47
Natural killer (NK) cell, 163
Naturally acquired active immunity, 187
Natural selection, 27
*Neisseria meningitidis*, 134
Neuron, 72–73
Neurotoxin, 79–80
Neurotransmitter, 74–79, 76
acetylcholine, 79
dopamine, 78
interaction of primary, 77
reuptake transporter, 75
serotonin, 74–78
Neutron, 29
Neutrophil, 163
News media, as source for emergency alert, 228
*Night of the Creeps*, 129
*Night of the Living Dead*, 85, 86
Nitrogen, 29
Nitrogen atomic structure, 29
NOAA weather radio alert, 230
Nonpolar molecule, 26, 33
Non Proliferation Treaty, 86
Nonspecific defense system, 161
Non-template strand, 91
Norepinephrine, 77
Normal microbiota, 161
Nucleic acid, 37
Nucleotide, 86, 87
Nucleus, 24, 25, 29

## O

$O_2$, 26
Occipital lobe, 66, 67, 70
Occupational Safety and Health Administration
        (OSHA), 151
OCD, 129
*Ophiocordyceps*, 126
*Ophiocordyceps unilateralis*, 126
Opportunistic pathogen, 148, 161, 186
Organelle, 24, 25
functions of, 24
Organic, 33
Organic macromolecules, 36–37
Organic molecules, 34
OSHA. *see* Occupational Safety and Health
        Administration
Outbreak, 147
Outdoor warning siren, 228
Oxygen, 29, 33
ATP and, 47–48
Oxygen atomic structure, 32

## P

Pain medicine, 229
Pandemic, 147
Pan flu plans, 201–203
Papua, New Guinea, 132
encephalitis and, 132
Paralysis, 79
Parasite, 120, 126–129
animals and, 128, 135
Parasitic protein, 129
Parasitism, 126
Parietal lobe, 66, 67
Parkinson's disease, prions and, 131
Pathogen, 114, 121
Pathogen control
personal protective equipment, 176–177
water and sewage, 175–176
Pathogenic bacteria, 117
Peer-review, 6, 14
Peer-reviewed journal, 6, 8
Penicillin, 182, 183
Periodic table, 28